THE CHALLENGE OF
GLOBAL
WARMING

THE CHALLENGE OF
GLOBAL
WARMING

Edited by

DEAN EDWIN ABRAHAMSON

ISLAND PRESS

Washington, D.C. □ Covelo, California

© 1989 Island Press

Every effort has been made to reproduce material presented in congressional testimony and at conferences in an accurate form. Some illustrations may have been omitted because of difficulties in reproduction or repetition of content.

Chapter 11 is reprinted from *Changing Climate* by Roger Revelle and Paul Waggoner, by permission of the National Academy of Sciences. *Changing Climate,* © 1983 by the National Academy of Sciences.

Library of Congress Cataloging-in-Publication Data.

The Challenge of global warming.

  Bibliography: p.
  Includes index.
M1. Global warming. 2. Climatic changes.
3. Greenhouse effect, Atmospheric. I. Abrahamson, Dean E.
QC981.8.G56C42   1989   551.6   89-1830
ISBN 0-933280-87-4
ISBN 0-933280-86-6 (pbk.)

Printed on recycled, acid-free paper
Manufactured in the United States of America
10   9   8   7   6   5   4

To Lisa, Eric, Tolli, and Hedinn,
whose generation inherits a planet badly served

ABOUT ISLAND PRESS

Island Press, a nonprofit organization, publishes, markets, and distributes the most advanced thinking on the conservation of our natural resources—books about soil, land, water, forests, wildlife, and hazardous and toxic wastes. These books are practical tools used by public officials, business and industry leaders, natural resource managers, and concerned citizens working to solve both local and global resource problems.

Founded in 1978, Island Press reorganized in 1984 to meet the increasing demand for substantive books on all resource-related issues. Island Press publishes and distributes under its own imprint and offers these services to other nonprofit organizations.

Funding to support Island Press is provided by The Mary Reynolds Babcock Foundation, The Ford Foundation, The George Gund Foundation, The William and Flora Hewlett Foundation, The Joyce Foundation, The J. M. Kaplan Fund, The John D. and Catherine T. MacArthur Foundation, The Andrew W. Mellon Foundation, Northwest Area Foundation, The Jessie Smith Noyes Foundation, The J. N. Pew, Jr. Charitable Trust, The Rockefeller Brothers Fund, and The Tides Foundation.

ABOUT THE NATURAL RESOURCES DEFENSE COUNCIL

The Natural Resources Defense Council, Inc. (NRDC) is a nonprofit membership organization dedicated to protecting America's natural resources and to improving the quality of the human environment. With offices in New York City, Washington, D.C. and San Francisco, and a full-time staff of lawyers, scientists, and environmental specialists, NRDC combines legal action, scientific research, and citizen education in a highly effective environmental protection program. NRDC has approximately 100,000 members and is supported by tax-deductible contributions.

NRDC's major environmental protection accomplishments have been in the areas of air and water pollution, energy policy and nuclear safety, control of toxic substances, conservation of natural resources, and international action to protect the global environment. In 1988, NRDC launched the Atmospheric Protection Initiative, a comprehensive advocacy campaign to advance solutions to the world-wide crisis of atmospheric pollution. The Atmospheric Protection Initiative combines the efforts of 20 NRDC scientists and attorneys to address the risks posed by global warming, ozone depletion, acid rain, and urban smog. The strategy of the Initiative includes legal action, legislative advocacy, public education, and the creation of projects demonstrating new methods and technologies, which can use energy more efficiently, and thereby help stop air pollution.

CONTENTS

PART III
GLOBAL WARMING: PHYSICAL IMPACTS

PART IV
THE GREENHOUSE GASES

PREFACE

Fossil fuel burning, deforestation, and the release of industrial chemicals are rapidly heating the earth to temperatures not experienced in human memory. Limiting global heating and climatic change is the central environmental challenge of our time.

This book summarizes the scientific aspects of the greenhouse effect and climatic change, explains why the issue is important, and shows that there are measures which, if implemented soon, can reduce the social, economic, environmental, and political impact of changing climate.

The presentation is intended for the nonscientist—the policy analyst, the legislative staff member, the advocate of environmental values, the student. Each part opens with a chapter written especially for this book by a recognized expert, followed by chapters, previously published elsewhere, which focus more narrowly on some aspect of climatic change. Considerable use has been made of recent congressional testimony, as it represents what is in many cases the only material written by leading scientists for a nonscientific audience.

The introduction by Senator Wirth and the chapters which open each of the five parts provide a comprehensive, albeit general, description of the issue from its grounding in basic science to effective policy responses. A lightly annotated bibliography of recent scientific and policy literature is included as an aid to the reader who needs access to primary sources.

This book would not have been possible without the help and support of Charles Savitt, Karen Berger, Nancy Seidule, and their colleagues at

Island Press. The Joyce Mertz-Gilmore Foundation provided generous financial support for my work which led to this book. John Firor, Rafe Pomerance, and George M. Woodwell, who number among those who have long striven to preserve some semblance of environmental integrity, contributed their chapters under extreme time pressure. My colleague Peter Ciborowski not only contributed the chapter reviewing the greenhouse gases but also provided able guidance and assistance throughout the project. Finally, I acknowledge the support, encouragement, and understanding given by my wife, Sigrun Stefansdottir.

DEAN EDWIN ABRAHAMSON

INTRODUCTION

Senator Timothy E. Wirth

In the United States, the year 1988 will be remembered for many historic events. The U.S. reached agreement on a major arms control treaty with the Soviet Union and the American people elected a new president. Yet, one can make a strong case that the most significant events were a string of natural disasters that took place in the summer of 1988.

Throughout the summer, drought gripped vast segments of the North American continent. The parched conditions that ravaged crops on the most productive soils on earth were matched by record low flows in the mighty Mississippi River. By August, explosive forest fires burst on millions of acres of land in the American west. And the heat wave that accompanied these events brought record temperatures to communities throughout the United States.

Meanwhile, another phenomenon was occurring. Scientific alarm about the buildup of the so-called "greenhouse gases" and the theory that these gases could lead to global climate change was sounded with new strength. A warning was trumpeted to public policymakers and the American people—and it was heard. The coverage of this issue by the American media was explosive.

Global warming is an issue that has been passed from the world's scientists to its public policymakers. That is not to say that incontrovertible scientific data have been collected and disseminated that makes policymaking easy. On the contrary, a number of scientific uncertainties continue to surround this issue, and those uncertainties will be discussed in this book, along with those findings with which scientists are satisfied. For policymakers, however, I believe there is sufficient evidence upon which a number of public policy initiatives can be based.

Public policy is often created to try to influence a desired outcome. Unquestionably, it makes good sense to do all that we can to retain the predictability of our climate. Climate is our ally. If we threaten that alliance, we could jeopardize the very climatic conditions upon which nations plan themselves. Global warming is a challenge as compelling and as imperative as nuclear arms control. Unfortunately, we cannot negotiate with the climate. Instead, the nations of the world must make choices, unilaterally and collectively, to adapt our behavior to maintain existing climatic conditions.

How do we respond to scientific evidence that the earth could warm by anywhere from 1.5 to 4.5°C by the middle of the next century? One can measure the development of the industrial age in the increased atmospheric concentrations of carbon dioxide (CO_2) from 280 parts per million to today's level of more than 345 parts per million. Likewise, we have developed chlorofluorocarbons (CFCs), created tropospheric ozone, and increased emissions of nitrous oxide and methane, which together comprise the other half of the greenhouse gases. We have learned that many of the tools we've used to power the unprecedented development of society could imperil our future.

To reduce the buildup of these gases and take command of the future, we must change course. Political leadership is needed now, at the highest levels, to halt the global experiment that is currently under way. The governments of both industrialized and developing nations must begin now to implement policies to deal with this enormous environmental problem.

Let me outline some of the first steps that we should begin taking to address the buildup of the greenhouse gases. These initiatives make good public policy sense regardless of whether the greenhouse theory proves to be correct. There are four broad areas where we can begin: international negotiations, energy policy, natural resource policy, and population policy.

In March of 1988, I joined 41 of my colleagues in urging President Reagan to raise the greenhouse issue first with General Secretary Gorbachev in Moscow and then at the Seven-Nation Economic Summit held in Toronto. Specifically, we urged President Reagan to call upon nations to begin the negotiations of a global climate convention.

International negotiations should begin immediately to organize international research and begin developing consensus among nations on strategies to reduce the risks of climate change. The initial goal should be to establish an international global climate convention by 1992.

International meetings should also begin now on new scientific evidence about the depletion of the earth's protective stratospheric ozone layer. The parties to the Montreal Protocol on Substances That Deplete the Ozone Layer should agree to reduce this threat altogether by setting a speedy timeline for the complete phase-out of ozone-depleting chemicals. This action would also eliminate the source of 15 to 20% of the greenhouse gases: CFCs.

Energy policies around the world must also be redirected to cut CO_2 and other trace gas emissions. The process of reducing emissions of greenhouse gases is going to be controversial and, no doubt, very painful for many nations. But if the global warming trends continue, the major industrialized nations will have to collectively alter the ways in which we organize society and conduct business in much more drastic ways.

The industrialized nations must also aggressively assist the nonindustrialized nations to develop sustainable, energy-efficient, vibrant economies. Economic development *is* consistent with protection of the earth's climate, and economic growth and energy efficiency can and must go hand in hand. In fact, done properly, economic development designed to be energy efficient, and to utilize new, less harmful sources of energy, will leave scarce economic resources available for other, nonenergy investments.

Here in the U.S., the oil price collapse of the 1980s has lulled the nation into complacency about energy policy when it is becoming clear that we need to minimize the by-products of fossil fuel combustion. Increased reliance on cheap, foreign oil has again questioned the future of our energy security. Imports continue to climb in this nation and, according to many energy analysts, could approach 50% of domestic consumption by the 1990s. In 1987, the U.S. paid $40 billion for its imported oil, 26% of the nation's mammoth merchandise trade deficit. Clearly, an energy policy that reduces our foreign energy dependence will improve our trade imbalance and enhance our national security.

The most ambitious forecasts estimate that the U.S. can save more than $150 billion on our annual energy bill by making energy-efficiency improvements in the transportation, industrial, building, and government sectors. The U.S. must make energy efficiency a top energy priority, as well as one of the primary strategies for addressing environmental concerns.

The major industrialized nations also should step up research and development programs for alternative sources of energy that will not contribute to emissions of greenhouse gases. Photovoltaic solar energy is one of the technologies that, again, has multiple benefits. Solar energy is environmentally benign and can be used to benefit the lives of citizens of the developing nations by providing flexible sources of electricity in crucial economic, health care, and other sectors, especially in rural areas. We can also bolster the world's research and development efforts on wind energy and other renewables.

One of the most controversial issues that must be a part of the broad search for solutions is nuclear power. The current generation of nuclear energy technologies is fraught with financial and safety problems, but I believe we should step up our research to determine if we can develop a new generation of passively safe, economical, nuclear power plants.

Nuclear power has little if any chance of playing an expanded role in the United States for the next decade. It is not "the" solution to this or any other problem. But we must not foreclose the possibility that safe and cost-effective nuclear power can be developed. And if it can, it is not a foregone conclusion that the public will embrace stacks of data about safety. Nonetheless, we should attempt to see if new nuclear technologies can in some way help retain the integrity of our atmosphere without jeopardizing our planet in other ways.

Sound natural resource policy demands that the developed and developing nations work together to halt the devastating rise in tropical deforestation, the other force driving increases in atmospheric CO_2 concentrations. More than 7.5 million hectares of closed forests and 3.8 million hectares of open forests are cleared in the tropics each year. Rapid rates of deforestation reduce carbon storage and increase the inventory of atmospheric CO_2.

For several years, conservationists have been urging the multilateral lending institutions to give much greater environmental scrutiny to loans to developing nations. All too often, these projects yield small economic benefits and very great environmental costs. International representatives to the multilateral lending institutions must exercise their responsibilities to ensure full consideration of the environmental impact of projects before approving loans to developing nations. Further, the United States and other lending countries should expedite the use of debt instruments in exchange for preservation of rain forests— such a policy is good for everyone. Our common goal should be sustainable economic development—and, again, we must recognize that good environmental policy is good economic policy.

We in the United States can set a good example by halting the dreadful ripping down of the Tongass National Forest in Alaska—the last great rain forest in North America. On this and other issues, our government should proudly match its deeds with its words.

We must also address the high rates of population growth in the world's poorer countries that contribute both to the rise in total world demand for energy and to the rapid destruction of tropical forests. A devastating cycle is occurring whereby growing rural populations in the developing nations are invading the remaining forests in search of land for food, commercial crops, and fuelwood for cooking and heating.

This is a disastrous strategy for meeting the needs of the rising Third World populations. Half of the world's people depend on biomass energy—principally fuelwood—and about 60% of these people, or about 1.5 billion people, are cutting wood faster than forests can grow back, creating a fuelwood deficit that is expected to double by the year 2000.

These developments are increasingly being strained by continued population growth. At current birth and death rates, the world population, now at 5 billion, will grow by at least 1 billion people every 10 years. Developing countries grow by 1 million people every 4½ days. Unless much greater progress is made to reduce birthrates, the world's population will have doubled to 10 billion by the year 2050, and some nations may double in size in the next 25 years.

This doubling of the world's population, however, need not be inevitable. If one looks at the fertility rates of 1960 and 1987 we can see that fertility declined between 35 and 75% in Singapore, Taiwan, South

Korea, Cuba, China, Chile, Brazil, Mexico, and Malaysia, and the list goes on. In most of these countries, social and economic changes, particularly improvements in the status of women, contributed to declines in desired family size.

Population growth is an enormously controversial and complicated issue, but it is also one of the single greatest challenges of the modern age. By avoiding this central question, as is largely the case today, we are postponing the inevitable. It is an issue that will have to be addressed forcefully, and the time to exert that leadership is now.

My last suggestion is for the leaders and scientists of the world to begin a campaign of public education. We are entering a new age, I believe, in which issues that strike to the very future of life on this planet will have to be addressed through open and informed debate.

Without recognition and action, we will continue on a course that is rapidly and fundamentally altering the composition of the atmosphere. By doing nothing we risk the planet's future. We *must* change directions. I hope that history will record that at the end of the twentieth century, mankind recognized and began to meet its greatest environmental, economic, and political challenge.

PART I

THE CHALLENGE OF GLOBAL WARMING

Chapter 1

GLOBAL WARMING: THE ISSUE, IMPACTS, RESPONSES

Dean Edwin Abrahamson

THE ISSUE

Humanity is conducting an unintended, uncontrolled, globally pervasive experiment whose ultimate consequences could be second only to
nuclear war. The Earth's atmosphere is being changed at an unprecedented rate by pollutants resulting from human activities, inefficient
and wasteful fossil fuel use and the effects of rapid population growth
in many regions. These changes are already having harmful consequences over many parts of the globe.
                    —*Toronto Conference statement, June 1988*

This analogy between the consequences of nuclear war and atmospheric
pollution was made not by idealistic, scientifically innocent environmentalists, but by the more than 300 policymakers and scientists from
46 countries, United Nations organizations, other international bodies,
and nongovernmental organizations who attended a major international

3

conference sponsored by the government of Canada. The Toronto Conference statement, included in full in Chapter 3, illustrates that it is now clearly within our power not only to alter the planet beyond comprehension within a few hours by using nuclear weapons, but also within a few decades by destroying the earth's life-support systems and radically changing climate by contamination of the air and water with the residuals of production.

Global climate is changing because of the buildup in the atmosphere of carbon dioxide (CO_2), methane, nitrous oxide (N_2O), the CFCs (powerful greenhouse gases as well as destroyers of stratospheric ozone), and other greenhouse gases produced by fossil fuel burning, by deforestation (discussed in detail in Chapter 5), and by producing food for the rapidly increasing global population.

The consequences of the global heating which would result if present release rates of the major greenhouse gases are maintained for only a few more decades would be catastrophic if our present scientific understanding is even approximately correct, and the resulting climatic change would be irreversible. (See the report of the 1987 Villach–Bellagio conferences in Chapter 7.) It is now known that a warming of several degrees, greater than previously experienced in human history, could occur within the next few decades—a time which is short compared with the lifespan of a tree or a man. Major changes will result in ecological, economic, and social systems as all are in delicate balance with their environments which in turn are dependent on climate. No one can now describe the precise nature of these changes—in part because of the technical demands of climate modeling and in part because of the impossibility of predicting the choices which will be made within the next 50 or so years—but change there will be. Although the crystal ball is too cloudy to reveal the details, the general course of climatic change is well understood (see Chapter 8).

Studies of past climate changes can tell us something about the response of forests. As the last ice age retreated, the earth warmed, slowly, for several thousand years. This change completely rearranged the face of the United States. Tree species that grew in Ohio and Michigan migrated far into Canada, and other tree species from the Deep South moved north into Ohio and Michigan.

If we allow emissions of CO_2 and the other greenhouse gases to continue unabated, the earth will warm five to ten times faster than it did during the retreat of the last ice age. Many trees cannot migrate much faster than they did as the earth slowly warmed following the last ice age. By the time a tree matures enough to produce seeds, the climate will be unfavorable for those seeds to take hold and produce the next generation. We thus could be heading toward a country almost devoid of young trees, a country given over to shrubs that tolerate a wide range of conditions—a sumac world.

As go the forests, so go the other species, animal and plant, supported by them. As Robert L. Peters suggests in Chapter 6, the most likely outcome is widespread extinction of species. The U.S. National Research Council has concluded: "It seems likely that the impacts of climatic change will fall most severely on immovable, and therefore inflexible, elements of both natural and man-made infrastructures. . . . National parks and biosphere reserves are usually established to preserve some asset of unique physical or biological importance, often depend on climatic factors, and cannot be easily removed or replaced."[1]

Each 1°C of global warming will shift temperature zones by about 100 miles. A continuation of present trends in the emission of CO_2 and the other greenhouse gases is expected to result in additional global heating of at least 2°C by the year 2030. Were this to happen, the climate upon which, for example, Yellowstone Park's ecosystem depends will have moved—the place called Yellowstone will be occupied by a different ecosystem. All the effort which has gone into the park from its creation through fighting the 1988 fire will have been in vain if global warming is not limited. Other parks, refuges, and wilderness will also be affected by warming, by changes in water balance, or by saltwater intrusion.

Continued global heating will also increase sea level by 1 to possibly 3 meters within the next hundred years (detailed in Chapter 12). This sea-level rise would be sufficient to inundate most salt marshes and coastal wetlands, change the character of the Everglades, push barrier islands further toward the present coast, and contaminate coastal aquifers. Reduced summer soil moisture would result in the loss of freshwater wetlands, reduced stream flows, and further lowering of aquifers. The consequences would include loss of wetlands habitat, reduced water quality, and increased concentrations of toxic wastes.

Coping with global heating and global climatic change may be the ultimate environmental challenge. If we fail to reduce emissions of the greenhouse gases, the climate change expected within the next few decades will be sufficient to jeopardize resources which much of the present environmental legislation is designed to protect.

The Brundtland Report details the challenge of making room for a world population of 8 to 14 billion people within the next century and a severalfold increase in world economic activity. A world with a doubled or tripled human population, with a severalfold increase in consumption, and with greenhouse gases, industrial pollutants, and other assaults on the environment proportional to those of today is not only virtually unimaginable, but impossible. If societies attempt a severalfold increase in economic activity described in the Brundtland Report using the present means of production, increasing emissions of greenhouse gases will have consequences similar to those of nuclear war. We have no alternative but to devise means of production which can provide the necessary goods and services to a growing population without causing irreversible biotic impoverishment.

THE GREENHOUSE EFFECT

Global warming has reached a level such that we can ascribe with a high degree of confidence a cause and effect relationship between the greenhouse effect and the observed warming.
—*James Hansen, NASA climatologist, 1988*

The greenhouse effect results from the buildup in the atmosphere of gases which absorb heat (long-wavelength infrared radiation), a topic considered in detail by Gordon MacDonald in Chapter 9. To maintain a constant average temperature, the earth must radiate heat to space. Greenhouse gases like CO_2 and methane absorb a part of this heat energy and reradiate it back to the surface of the planet, thereby effectively trapping it in the lower atmosphere. This process raises the temperature of the atmosphere near the earth, which in turn raises sea level, increases evaporation and precipitation to affect global cloud cover, and thereby alters the distribution of climate across the surface of the planet. An increase in average global temperature of 0.7°C has already been measured, and present rates of emission of greenhouse gases are committing the earth to an additional warming of about 0.3°C per decade.

It is because they trap heat like glass in a greenhouse that these gases received their name—the greenhouse gases. They are vitally important for life. Venus, with an atmosphere rich in CO_2, has a greenhouse effect of between 400 and 500°C, and Mars, with its thin atmosphere, only a few degrees. Earth has a natural greenhouse effect due to the presence of CO_2 and water vapor. If Earth's atmosphere did not contain these gases, its temperature would be 33°C lower. The natural level of these gases make life as we know it possible.

Increasing them beyond today's level will cause the climate to diverge markedly from its present state. Climatic zones shift by about 100 miles for each 1°C of global warming. Sea level will rise all over the earth. Major weather patterns—for example, the tropical monsoons and jet streams—are altered. A warming of about 4°C, to which we may be committed in less than 50 years, would result in an ice-free Arctic Ocean which will not only devastate arctic ecosystems but will further change climate and weather. A warming of this extent could also begin an irreversible disintegration of the West Antarctic ice sheets which would result, within a few hundred years, in a further rise in sea level by at least 6 meters.

ABOUT THE GREENHOUSE GASES

Carbon dioxide, the single most important greenhouse gas, accounts for about half of the warming that has been experienced as a result of past emissions and also for half of the projected future warming. The present concentration is now about 350 parts per million (ppm) and is increasing about 0.4% (1.5 ppm) per year, retaining an additional 3 billion tons (10^9 tons or GT) of carbon per year. For a very long period of time before the industrial age, the atmospheric concentration was essentially constant. Beginning about 1850, however, the CO_2 level began to rise, due in large part to the industrial burning of first coal and then oil and natural gas. The atmospheric CO_2 level in 1850 was probably about 270 ppm; thus it has already increased by about 30%. Human activities are now causing at least 7 billion tons of carbon,[2] as carbon dioxide, to be released into the atmosphere. Fossil fuel use releases about 6 billion tons per year to the atmosphere. In addition, the clearing and burning of forests causes the pool of carbon which has been tied up in trees to be released. Deforestation is in part the result of population expansion in the tropics and the use of land for agriculture and wood for energy. The desire of developed countries for inexpensive meat and forest products is another major contributor to deforestation. Each year a land area the size of Belgium is deforested, resulting in the transfer from forests to the atmosphere of between 1 and 3 billion tons of carbon per year, as detailed by George M. Woodwell in Chapter 5.

If present trends in the release of CO_2 to the atmosphere continue, the preindustrial level of CO_2 will increase another 30% in 50 years, and then double about the year 2100. In principle, there are no obvious physical constraints to limit this rise. The remaining amounts of oil and gas are quite small, but the amount of coal that is subject to exploitation is essentially unlimited. Economics and policy on emission of greenhouse gases will govern how rapidly CO_2 builds up before either humankind or nature acts to stabilize the atmospheric burden of greenhouse gases.

In addition to CO_2, another 20 or so greenhouse gases have been identified (detailed by Peter Ciborowski in Chapter 14). At present the most important are methane (CH_4), the chlorofluorocarbons CFC-11 and CFC-12, nitrous oxide (N_2O), and ozone (O_3) in the lower atmosphere (see Chapter 15).[3] Taken together, these other greenhouse gases are responsible for at least as much global warming as is CO_2, and perhaps more. At present methane is the most important of these gases, followed closely by CFC-12.

Methane is produced in flooded fields and waterlogged soils, in rice paddies, in the guts of cattle and other fauna, in landfills, and in coal seams. It is also released as a result of forest clearing, venting in association with oil production, and leakage from natural gas pipelines. Atmospheric methane concentration is now increasing at about 1% per year (see Chapter 17). CFC-12 is primarily used as the working fluid in refrigerators and air conditioners. CFC-11 is used mainly in plastic urethane foams. The CFCs in general are used in spray cans, forming plastic foams, and as solvents. (See Chapter 20.) Nitrous oxide is released as a result of coal combustion and in the breakdown of agricultural fertilizers. Tropospheric ozone is photochemically produced in the atmosphere as a result of the release of methane, carbon monoxide, and other hydrocarbons, due largely to emissions from fossil fuel burning.

Atmospheric concentrations of the principal non-CO_2 greenhouse gases are increasing from between 0.3% per year in the case of nitrous oxide to 5% per year in the case of CFC-11 and CFC-12. (See Chapter 16.) Tropospheric methane and ozone are increasing in concentration about 1% annually. All these rates of increase, if continued or only marginally reduced, will result in climatically significant atmospheric accumulations of these gases.

CLIMATE SENSITIVITY

As CO_2 and the other greenhouse gases accumulate in the atmosphere, the temperature near the surface of the earth will rise. Climate modelers and other atmospheric scientists have performed elaborate calculations to determine how much it might warm (see Chapter 8). The estimates are usually expressed in degrees centigrade of mean global temperature change—the rise of surface temperature averaged across the globe. They are typically given for a standard experiment—a doubling of the amount of atmospheric CO_2, which is a measure of how sensitive the climate is to the greenhouse gases. The best of the present climate models show that if atmospheric CO_2 were to double, the average global temperature would rise by between 3.5 and 4.5°C. By way of comparison, the typical natural variation of mean global temperature over periods of 100 to 200 years has been at most 0.5 to 1°C—smaller than the expected rise in mean global temperature for doubled CO_2 by at least a factor of 4. The warming would be two or three times the global mean in high latitudes and less than the global mean near the equator.

THE OCEAN THERMAL DELAY

It will require a much longer period of time to warm the oceans than it will to warm the atmosphere. Thus there is a delay of at least a decade and perhaps half a century between the time that greenhouse gases are added to the atmosphere and the time that the full warming effect of these cases can be measured. It is thus necessary to distinguish between the eventual global heating which will result from any given amount of greenhouse gases added to the atmosphere—the equilibrium warming—and the heating which is experienced at any point in time—the transient warming.

It is essential to note that the greater the warming from the greenhouse gases, the longer the ocean thermal delay. If there is a global warming of only 1.5°C for doubled atmospheric CO_2, the delay may be only about 15 years; but if doubled CO_2 results in 4°C warming, the delay may be from 50 to 100 years. As a consequence, the global warming which has been observed to date is consistent with either a low climate sensitivity to increased greenhouse gases, accompanied by a short ocean thermal delay, or with a high climate sensitivity which would be accompanied by a long ocean thermal delay.

TODAY'S GLOBAL WARMING

Measurements have shown that the earth's surface temperature has increased by between 0.5 and 0.7°C since 1860. Current climatic models show that a transient warming of between 0.5 and 1.0°C from greenhouse gases added to the atmosphere during the last century should have occurred. If we consider the equilibrium warming that should have resulted from these emissions, however, the situation is different, as past emissions should have committed the planet to an equilibrium surface warming of between 1 and 2.4°C. The difference between the calculated equilibrium, or committed warming, and the observed warming—0.5 to 1.7°C—is the unrealized warming from past emissions. This warming will be expressed over the next few decades regardless of what we do about future releases of the greenhouse gases.

At the present rate of emissions of the greenhouse gases, we are committing the earth to an additional warming of at least 0.15°C and perhaps as much as 0.5°C each decade. If the present rates of growth in emissions were to continue, by the year 2030 we would be committed to a mean global warming of at least 3°C and perhaps as much as 5°C. (See the 1985 Villach Conference statement in Chapter 4.) Over a century this figure could reach 5 to 10°C. This would change the earth, in less than a single human lifetime, to a climatic regime not experienced for millions of years.

CRITICAL PARAMETERS AND UNCERTAINTY

There is considerable uncertainty about several of the most critical parameters influencing global climate change. Some of the uncertainties are because of incomplete scientific understanding; others arise because we cannot know what future releases of greenhouse gases will be deemed politically acceptable. Thus the equilibrium climate sensitivity to doubled CO_2 concentrations could be as low as 1.5°C or as high as 5.5°C. Most contemporary models give results at the upper end of this range. The warming resulting from the other greenhouse gases might be as small as 50%, or as large as 300%, as the warming from CO_2 alone. The ocean thermal delay could be as short as 10 years or as long as a century.

Global sea level will rise as the oceans warm and their waters expand and as ice sheets and glaciers melt. Within the next century sea level could rise as little as 20 centimeters or as much as 2 meters, and the Ross and Filcher-Ronne ice shelves in West Antarctica could be thinned to the point that the West Antarctic ice sheet could collapse. This would add a further 6-meter rise in sea level over the next two to four centuries. (The causes and effects of sea level rise are summarized by James G. Titus in Chapter 12.)

Finally, there is considerable uncertainty about the amount of CO_2 that might accumulate in the atmosphere. It could be held to a value as low as 400 to 500 ppm if vigorous energy conservation measures are taken, primary energy sources are shifted from the fossil fuels, and reforestation replaces deforestation. It all depends on energy and emissions policies (see Chapter 19). Should the use of coal increase significantly within slightly more than a century, the atmospheric level of CO_2 could increase to well over 800 ppm. The present situation is not encouraging. The U.S. Department of Energy expects a 40% increase in coal consumption between 1985 and 2000—in spite of a recognition that coal produces more greenhouse gases than any other current energy source.[4] The energy industry has responded to environmental concerns about acid precipitation by developing technologies which permit the use of coal while in large part avoiding or reducing conventional air pollution. This new technology does nothing, however, to limit emissions of carbon dioxide and the other greenhouse gases. Unless economic means are devised to remove the CO_2 from coal effluents, there can be no clean coal.

Two of these fundamental parameters are of critical importance: the atmospheric sensitivity to increased concentrations of greenhouse gases (a matter of science) and the future releases of these gases (a matter of politics). If the atmosphere warms only about 1.5°C for doubled concentrations of CO_2 or the equivalent in other greenhouse gases, then not only is the earth committed as a result of past emissions to a relatively small warming but also, as the ocean thermal delay would be short, the measured warming would be very close to the total expected equilibrium warming. If, on the other hand, the atmosphere heats 3 to 5°C for the same greenhouse gas loading (as is indicated by the most sophisticated climate modeling) then not only are we already committed to a major climate change but, as the ocean thermal delay would then be many decades to perhaps a century, the heating to which we are irreversibly committed at any given time would be much larger than the measured warming.

The wide range of climate sensitivities means that a business as usual approach would result, by about the year 2070, in a rise in mean global temperature of as low as 3°C or as high as about 6°C. It will be difficult to adapt to even the most optimistic case of a 2°C average global heating, but it will be something we must accept and learn to live with.

POSITIVE FEEDBACKS

Carbon dioxide and methane production and release will increase as global temperature rises, as described in Chapter 5. As the earth warms and the atmospheric concentration of CO_2 increases, the CO_2 removed from the atmosphere by green plants through photosynthesis increases. But even more CO_2 is added to the atmosphere because biotic respiration, which produces CO_2, increases more rapidly than does carbon fixation through photosynthesis. Methane production in waterlogged soils, bogs, marshes, and rice paddies, which accounts for as much as 40% of all methane emissions into the atmosphere, likewise increases with temperature. In Chapter 17, Donald R. Blake points out that increasing temperature can have an "incredible effect" on methane production—a 1°C warming may increase methane production by as much as 20 or 30%. These temperature-sensitive releases of CO_2 and methane could be very large and add significantly to global warming.

CHANGES IN OTHER PHYSICAL PARAMETERS

Global warming will increase the frequencies of extreme weather events. NASA climatologist James Hansen, for example, has shown that with global warming resulting from doubled CO_2 or its equivalent, the frequency of days per year with temperature exceeding 100°F would increase from 1 to 12 days for Washington, D.C., 3 to 21 for Omaha, Nebraska, and 19 to 78 for Dallas, Texas.

Global heating also alters other climate parameters. Since the earth is warming more near its poles than near the equator, greenhouse warming results in a smaller temperature difference between the poles and the equator. It is this temperature difference which drives major weather systems and ocean currents. A warmer earth thus will have less vigorous ocean currents, less vigorous continental weather systems, more intense tropical storms, and altered jet streams and monsoonal circulation in the tropics.

A warmer world will be a wetter world, too, as evaporation increases with temperature and what goes up must come down. The distribution of precipitation, both geographically and by season, will change, although the global climate models are less reliable regarding regional changes in precipitation than they are for the temperature response. Precipitation should increase in low-latitude regions (0 to about 30 degrees) which have heavy rainfall today, and during the winter in high latitudes (about 60 to 90 degrees). Mid-latitudes, particularly in continental interiors, may have reduced summer precipitation.

Many of the world's major agricultural regions, including the U.S. grain belt, the Canadian Prairie Provinces, the Ukraine, and northern China, may experience much reduced summer soil moisture because of earlier snowmelt, reduced precipitation, and increased water loss through evaporation and transpiration. Syukuro Manabe has shown (Chapter 10) that the summer soil moisture in much of the U.S. Midwest could be reduced by 40 or 50% with a doubling of atmospheric CO_2.

Runoff and aquifer recharge rates should decrease by even more than available soil moisture—as was evident during the 1988 U.S. drought when temperatures were higher, and precipitation lower, than normal. As a consequence stream and river flows were much reduced, potholes and wetlands went dry, and aquifer levels fell because of increased irrigation and decreased recharge. These changes led in turn to reduced waterfowl populations, water rationing, stranded river barges, reduced electric power generating capacity, and decreased water quality as the capacity of sewage treatment plants is based on downstream dilution associated with customary river flows. A U.S. National Academy of Sciences study (see Chapter 11) shows that annual flow in several

Western and Southwestern U.S. rivers would be expected to decrease by from 40 to 75% with a temperature increase of only 2°C and a reduction in annual precipitation by only 10%—climate changes to which we are already committed.

IMPACTS

Not since the abrupt end of glacial climates a little over 10,000 years ago have temperatures changed as much, or so rapidly. The next century may therefore see large impacts on the human economy, with the first signs already upon us. . . . I have no doubt that we are discussing the central environmental problem of our times.
—*F. Kenneth Hare, Chairman of Canada's Climate Planning Board, 1988*

Natural ecosystems are in delicate balance with their environments and climates. The impacts of climatic change become more severe with increases in both the magnitude and the rate of change in average global temperature. Were climate changes to be small, and to occur sufficiently slowly, they could be relatively nondisruptive. The 1987 Villach Conference experts concluded that minimization of the disruption resulting from global warming would require rates of change no greater than 0.1°C per decade. But climatic change is unlikely to be slow, and at present it may be three times this rate (see Chapter 8). If our understanding of the forestry response is even remotely correct, the response of the unmanaged biosphere will be characterized by widespread species extinction and a general impoverishment of natural systems.

ENVIRONMENTAL VALUES

Activities and resources which much of the present U.S. environmental legislation is designed to protect are in jeopardy as a result of climate warming. In the Midwest and Far West a drier climate will affect the viability of forests at the margins of present steppe climates. Reduced lake levels and stream flows will affect freshwater ecosystems and make the preservation of endangered aquatic species difficult. They will also make it difficult to adhere to present water quality standards, which will need to be tightened with less downstream dilution of water-borne pollutants. In the north, warming and drying will threaten peat bogs, marshes, prairie potholes, and the species dependent upon them for breeding grounds or habitat. On the coasts, the rise in sea level will inundate most salt marshes and coastal wetlands, push barrier islands further toward the present coast, contaminate coastal aquifers, and affect a host of ecological systems (see Chapter 12). Change is the critical concept. Virtually every ecological system will be impacted, changed, and stressed. Areas that are being preserved for their special biological characteristics will lose their unique qualities as climate and vegetation change, and then possibly their protected status as well. The preservation ethic may well be rendered irrelevant in a world of continuous ecological change.

As climate continues to warm, the changes will become more dramatic. At some point in the next century, the floating ice pack in the Arctic Basin will melt. Not only will this radically impact the ecology and physical environment of the high latitudes, it will also alter the boundary conditions governing the climate. It is not clear how climate will respond to this change, but the physical effect will be enormous.

One must go back in time 5 to 15 million years to the late Tertiary to find a time that was 3 or 4°C warmer than now. During periods in which there was no permanent pack ice in the Arctic, climatic and vegetational regions and boundaries were displaced as much as 1,000 or 2,000 kilometers north of their present position (a displacement which we may replicate during the next 100 years). Intensely arid conditions prevailed in the United States from the Dakotas and Missouri to Alabama and throughout central and southern Africa. Because of the scale of the change, we understand very little of how the climate system might evolve, but it is likely to be so qualitatively distant from the present climate as to make wholesale upheaval in the natural environment a certainty.

It is difficult to see how such a change would not undo much of the past and present efforts at conservation and environmental preservation. The ecological or environmental resources that would be affected cut across the entire spectrum of environmental concerns: endangered species; habitat preservation; coastal zone protection; all legislation affecting inland rivers and lakes, dams, and water diversion; groundwater protection; wind-driven soil erosion and desertification (and hence western federal land management and reclamation); air quality; marine mammals; wilderness and other unique areas; natural parks; forestry and fisheries.

ECONOMIC ACTIVITIES

Economic activities that are sensitive to temperature and moisture—for example, agriculture—will be affected. It is impossible to predict with any degree of confidence the possible impact on yields, given the many structural and technological changes which the agricultural sector is presently undergoing. However, present agricultural sensitivity to climate—as evidenced in yield responses to heat waves and drought—shows large adverse impacts. As an example, a senior U.S. climatologist recently suggested that the impact on U.S. agriculture would be so great that by 2050 the United States could be importing wheat from the Soviet Union.

The effects of sea level rise are perhaps the best understood. sea level rise will inundate populated and unpopulated areas, flood harbors, erode beaches and cliffs, possibly break up barrier islands, lead to the intrusion of saltwater into drinking water supplies, and destroy coastal wetlands and the fisheries which depend upon them. Higher sea levels will also lead to higher storm surges and more damage from hurricanes. The number of hurricanes itself will probably increase, as will their destructive force, leading to additional economic losses.

On the basis of admittedly incomplete evidence, then, it appears that there is a reasonable chance of negative impacts for the following: grain, vegetable, and animal agriculture; fiber production; any activities dependent upon in-stream river flows (for instance, electricity and synfuels production, urban water uses, barge transport, and agricultural irrigation); many forms of recreation; coastal development, fisheries, and settlement patterns; and all forest-related industries. Water scarcity in Southern California, as well as fire incidence, would be of special concern if the 5°C scenario comes to pass. This concern stems from the dense pattern of regional population, the size of which makes any major impact—including outward migration of people—a significant national problem. The National Academy of Sciences suggests reductions in stream flows in the Colorado River and in the rivers of Northern California of as much as 40 to 70% and considerable stress between anticipated demand and supply. From where will replacement water come? What will be the costs of movement? Can it be moved? And if not, what then?

EVIDENCE FROM SUMMER 1988

North America experienced high spring and summer temperatures and drought during 1988—the conditions one expects with modest global warming. The news headlines summarize impacts better than any theoretical study and indicate in a convincing way what will be commonplace within a few years. Here is a sample:

- Worst forest fires of the century
- Water rationing
- Neighbors encouraged to report forbidden water uses to authorities
- Regional tensions associated with falling water levels in headwater reservoirs
- Downstream fish kills and lower water quality as sewage treatment discharges are undiluted
- Operation of power plants curtailed because of insufficient cooling water
- Crop failures
- Insurance companies attempting to default on drought insurance policies
- Increased use of groundwater irrigation
- Aquifer levels falling as recharge decreases and withdrawals increase
- Barge traffic in chaos
- Reduced growth in GNP attributed to drought
- Billions authorized in drought aid
- Lack of seasonal jobs depriving migrant workers of income
- Economic effects rippling through society

IMPLICATIONS FOR ADAPTATION

> Science Finds,
> Industry Applies,
> Man Conforms
> —*Motto of 1933 Chicago World's Fair*

Much of global warming policy analysis has been concerned with whether to pursue policies of adaptation or those of prevention. Heroic attempts have been made to show that adaptation can be painless, or nearly so, and that no significant changes are necessary in the way we do business. This approach fits with the times, still characterized by the philosophy so tersely summarized in the motto of the 1933 World's Fair. Even informed scientists, aware of the potentially devastating impact of rapid climate change on natural and managed systems, can convince themselves that trend is destiny and that conforming is the only option. (See, for example, the lead paragraph in Chapter 13.)

CONVENTIONAL ASSUMPTIONS

Some of the early global warming policy analyses seemed to show that coping would not be all that difficult and growth itself would provide the solution.[5] The 1983 U.S. National Academy of Sciences analysis, for example, concluded: "The foreseeable consequences of climate change are no cause for alarm on a global scale but could prove to be exceedingly bad news for particular parts of the world. Generally, the more well-to-do countries can take in stride what may prove to be a reduction (probably not noticeable as such) by a few percent in living standards that will likely be greater per capita by more than 100% over today's."[6] This view—that more growth will provide the solution to the problems created by growth—is still heard.

Other conventional analyses have reached similar conclusions— which is not surprising in that they proceed from identical, although usually unstated, assumptions: (1) that climate is moving from that which we now know to some new equilibrium state which can be described in as much detail as is necessary; (2) that global temperature will increase very slowly, changing climate by only small amounts in periods of time characteristic of the life of individuals or of capital, and without abrupt or unexpected events. It is now becoming generally recognized that neither assumption is valid.

It is not surprising that global warming policy has been slow to evolve. If it appeared that adaptation might be inappropriate, or impossible, then difficult decisions involving growth, choice, and freedoms would be unavoidable. These concerns, together with the recognition that global warming is, and will continue to be, a direct consequence of our most fundamental activities—the production of energy and food— have led to a powerful need *not* to know that controlling global warming is both essential and very difficult. It is unpleasant for the members of a technocratic culture to realize that there is no easy technical fix.

AN EMERGING CONSENSUS

The scientific community is rarely unanimous on an issue of wide public importance; its business is, after all, to question, review, examine, test, and argue in search of new insights and new data. But on the issue of climatic change there is surprising unanimity that it does pose an extremely serious threat to society and that policy responses are necessary.

The scientific consensus has now been expressed in a variety of ways including formal reviews by learned societies, by scientific conferences and workshops (see, for example, the statement from the 1985 Villach Conference in Chapter 4), by conferences of policymakers (see the Toronto Conference statement in Chapter 3), and, not least, through the testimony at congressional hearings. As a result of these activities by scientists, the alarming and largely unexpected discovery of stratospheric ozone depletion, and the recognition that the earth has become warmer than it has been for centuries, global warming is now moving rapidly onto the political agenda (summarized by Rafe Pomerance in Chapter 18), its importance is being recognized by influential political leaders (see, for example, Senator Wirth's introduction to this book), and detailed recommendations for policy changes are being formulated (see Chapters 3, 19 and 21).

As indicated in recent congressional testimony by the leading greenhouse effect researchers, and by statements of conferences of scientists and policymakers, more and more stress is being laid upon the need to reduce the rate and magnitude of global warming—and to do it quickly.

THE REQUIREMENTS FOR ADAPTATION

Societies must, one way or other, adapt to the climatic change which is unavoidable because of greenhouse gases already added to the atmosphere and those certain to be added while policy responses are debated, decided, and implemented. Yet some analysts still claim that it is either impossible or undesirable to limit climatic change and that societies can cope with the anticipated climate change through planned adaptation and in response to market forces—in essence, they argue that adaptation ought to constitute the response of society to climatic change, both inadvertent and that which we can yet avoid, and, therefore, that no action to limit emissions ought to be taken.[7] Whether planned adaptation is in fact possible depends on how rapidly the climate is changing. If the climate is changing so fast that conditions on which adaptive responses—a dike, a dam, an investment in agricultural infrastructure—depend, those investments will not be made or are likely to be ineffective.

Several important characteristics of the adaptive response are of note. First, the characteristic times of national capital stocks are typically very long—several decades. The time required for major changes in technological systems varies between 20 and 50 years and is probably between 30 and 50 years for water projects. Thus changes in climate cannot be excessively rapid or involve successive closely spaced regional climate changes. If the climate changes too rapidly, the usefulness of expensive interventions may be significantly less than is typically required in public investment decisions, rendering them unworkable. A rate of warming of 0.5 to 1°C per decade will soon prevail in middle and high latitudes in the absence of policy changes to reduce emissions of greenhouse gases. These rates of change are certainly too rapid to be economically workable.

Moreover, the adaptive response requires relatively certain projections of long-term *local* trends in precipitation, drought frequency, wind intensity, and the like expected during the physical or economic lifetime of capital investments. If expensive investments in adaptation are to be made, there must be a reasonable expectation for success. At present we do not have sufficiently reliable projections, nor are we likely to have them soon. The assessment of the U.S. National Academy of Sciences is typical: "We have little confidence in predictions about many details of the forecasts: local changes by city or state, exact shifts of desert regions or extent of monsoons, changes in river flows, or overall economic impacts."[8] Thus, for all practical purposes, most or all major *planned* changes in technological or other capital-intensive systems probably involve time scales which are longer than the time scales of economically significant climate changes and require information that we cannot even come close to providing at this time.

There are, of course, traditional *nonplanned* ways by which society often adapts to environmental change: changed soil or snow management practices in agriculture; water use controls; strengthened drought relief programs; enhanced preparedness for floods or hurricanes. But few of these short-term adaptations are well matched to climate changes that seem likely over the next 50 to 100 years. The new biotechnologies, which some claim are likely to be powerful forces of adaptation (and have particularly short lead times), may be an exception, although their availability has yet to be demonstrated and, like nuclear power, their political acceptability may be in doubt.

The dilemma here is simple. In the past we were able to adapt through these means because climate was essentially stable. Past climate changed little during the lifespans of trees or humans, human migrations involved far fewer people and fewer political barriers to movement, and plant or animal migrations were not blocked by development. We are now facing the imminent prospect of a climatic change, greater than any previously experienced, taking place in a time that is quite short compared with economic, social, or ecological time and in a world already stressed severely by rapid population growth and resource scarcity.

Beyond the year 2050 we enter a period that constitutes the very long term to the contemporary policy process. If by that time the mean global temperature were to rise 3 to 5°C, we could be committed to a far larger warming—probably on the order of 6 to 10°C. The climatic conditions that might be associated with such a warming are, with few exceptions, pure mystery. Certainly they cannot be outlined with the specificity necessary, for example, for planning major water projects. We simply do not know what might happen if it were to warm this much (see Chapter 13). Today's climate models have little, if any, validity for such extreme warming. Any long-term continuation of present business as usual policies will probably create a climate about which essentially nothing is known, for which planning is not possible, and to which preceding adaptations may be completely mismatched. There can be no planned adaptation under these conditions.

WINNERS AND LOSERS

Given the prospects for the adaptive response, it is no longer justifiable to delay action by citing the consensus about the details of future climate change, the need for regionally specific characterizations of future climates, or arguments about possible "winners" and "losers." The expected magnitude and rate of climate change led the participants at the Toronto Conference (Chapter 3) to conclude that no country will be a net winner. There can be only losers.

STABILIZING GLOBAL CLIMATE

Does the educated citizen know he is only a cog in an ecological mechanism? That if he will work with that mechanism his mental wealth and his material wealth can expand indefinitely? But that if he refuses to work with it, it will ultimately grind him to dust? If education does not teach us these things, then what is education for?
—*Aldo Leopold,* A Sand County Almanac

Eventual stabilization of global climate requires that the amount of greenhouse gases released into the atmosphere must be reduced to levels no greater than the rate at which these gases are removed by natural processes. The most important greenhouse gases are CO_2, methane, CFC-11 and CFC-12, tropospheric ozone, and nitrous oxide. These gases must be controlled.

Control can be achieved by some combination of measures that would remove greenhouse gases from the atmosphere or reduce the emissions of CO_2 and the other greenhouse gases. A technical mega-fix, inducing global cooling to offset the greenhouse heating, has also been suggested.

COOLING THE ATMOSPHERE

Programs which would induce just enough global cooling to offset undesired global warming would be the most desirable remedy using conventional criteria. They would involve massive government programs and therefore please planners and bureaucrats; they would be high-tech solutions and therefore please industry and its minions; they would provide employment for large numbers of scientists, engineers, and analysts; and they would free us all from the responsibility of examining the implications of our consumption.

Stratospheric dust and aerosols cool the atmosphere by blocking solar radiation. If introduced in sufficient quantities, these substances can offset global warming—presumably making it possible to continue the activities which release greenhouse gases without suffering the climatic consequences. Particulate matter could be injected into the stratosphere by using nuclear devices exploded near the earth's surface or by using aircraft or rocket systems. The orbiting of large reflecting plastic sheets to reduce the amount of sunlight reaching the atmosphere has also been proposed.

But Stephen Schneider, a leading climatologist from the National Center for Atmospheric Research, has noted: "This solution suffers from the immediate and obvious flaw that if there is admitted uncertainty associated with predicting the inadvertent consequences of human activities, then likewise substantial uncertainty surrounds any deliberate climatic modification. Thus, it is quite possible that the inadvertent change might be overestimated by our computer models and the advertent change underestimated, in which case our intervention would be a 'cure worse than the disease.' "[9]

REMOVING CO_2 FROM THE ATMOSPHERE

Carbon dioxide can be transferred from the atmosphere and stored in standing biomass through reforestation. Ecologist George Woodwell (Chapter 5) estimates that between 1 and 2 million square kilometers of newly planted trees would remove 1 GT of carbon annually.[10] That rate of storage would continue for 40 to 60 years until the forest reached stability and total respiration equaled gross photosynthesis. In fall 1988 it was announced that a new 180 MW(e) coal-fired power plant project would include reforesting 385 square miles in Central America. A new forest of that size, if permitted to survive, would remove from the atmosphere the CO_2 emissions from the plant over its expected 30-year lifetime.

Reforestation, while not sufficient by itself, could be part of the strategy to control climatic change, as must decreasing, and eventually stopping, deforestation. It is clear that although it is impractical to think in terms of solving the CO_2 problem through forestry alone, reforestation could play a small part and, of course, bring other benefits as well.[11]

Some have proposed that carbon could be fixed by organisms that live at the oceans' surface and then subsequently die and sink to the deep oceans. Even if there were no other arguments against the deliberate eutrophication of the oceans, the vast quantity of energy which would be required to produce and distribute the necessary fertilizer renders this scheme impractical.

REMOVING OTHER GREENHOUSE GASES FROM THE ATMOSPHERE

Fantastic schemes, including the use of powerful lasers to blast apart CFCs in the atmosphere, have been proposed.[12] No credible method has been described for removal of other greenhouse gases—other than to wait for natural processes to operate.

REDUCING CO_2 RELEASES

Carbon dioxide is the single most important greenhouse gas. Thus any attempt to limit future global warming would have to address CO_2 emissions. Five to six billion tons of carbon are now being released to the atmosphere per year from fossil fuel use and an unknown amount from deforestation and from increased biotic respiration resulting from the global warming which has already taken place. While the total carbon dioxide releases are not well known, we do know that the atmospheric concentration of CO_2 is increasing by 0.4% per year and that about 3 billion tons per year is being retained in the atmosphere annually. The rest of the CO_2 is being removed, presumably into the oceans.

Resolving the global carbon budget is of great scientific importance. Indeed, some analysts insist that the development of carbon emission policies must await a complete understanding of the global carbon budget. When we focus on what we do know, instead of what we do not know, however, it is clear that stabilizing atmospheric CO_2 levels requires that annual CO_2 releases be reduced by an amount *at least* equal to the 3 billion tons of carbon now being retained in the atmosphere—or at least 50% of present fossil fuel use. It is entirely possible that further research on the greenhouse gases will show that the stabilization of global temperature will require greater reductions in CO_2 emissions. From the perspective of policy, however, this matters little at present.

Present actions taken to reduce emissions by 50% would in no way preclude additional actions to further reduce emissions if in the future we find that such reductions are necessary.

There are several means through which CO_2 emissions might be reduced: scrubbing CO_2 from stacks; switching to fuels which produce less CO_2; and increasing energy end-use efficiency so that energy production can be reduced.

Scrubbing CO_2 from Stacks: Technology for collection of CO_2 from stack gases has been developed in response to the market for CO_2 used to stimulate secondary recovery of oil. Carbon dioxide scrubbing would, at best, be applicable to large fossil fuel burning facilities, such as electric power plants. These might account for 30% of the U.S. annual CO_2 production. It has been estimated that 90% removal of CO_2 would increase power plant investment capital by 70 to 150% and increase the production cost of electricity by 60 to 100%.[13] There is a further advantage to incorporating CO_2 scrubbers in fossil-fueled power plants, as recovery of more than 99% of the sulfur contained in the coal is required prior to CO_2 scrubbing.[14]

These proposals assume that the recovered CO_2 would be sequestered in the deep oceans, in spent oil and gas wells, or in excavated salt caverns, but the extent to which this strategy would contain the CO_2 is uncertain. Gregg Marland, a scientist at the Institute for Energy Analysis in Oak Ridge, Tennessee, has reviewed CO_2 scrubbers and other proposed technical fixes. He concludes: "We will have to look long and hard to conclude that practical collection and disposal of CO_2 is a reasonable possibility for dealing with the atmospheric increase."[15] It is likely, however, that some CO_2 removal will be proposed for the highly industrialized countries, and any new fossil-fueled power station should include CO_2 control—either scrubbers at the plant or adequate reforestation—as an integral part of the power plant.

Fuel Switching: The fossil fuels differ greatly in the amount of CO_2 they produce for the same net energy. Burning coal, for example, produces nearly twice as much CO_2 per unit energy as does burning natural gas. Electricity generated from coal produces more than five times as much CO_2 per unit energy as does heating directly with natural gas.[16] Thus, for example, a typical hot water heater using coal-generated electricity produces between three and four times as much global warming as does a gas-fired heater. Replacing coal and oil with natural gas will result in significant reductions in releases of CO_2.

Even more effective would be the replacement of coal-fired generating capacity with solar or other energy systems which produce no CO_2. In principle, all fossil fuel use could be eliminated and the world's energy needs met with some combination of primary energy sources which do not require the release of CO_2: the renewables, primarily solar using biomass and photoelectric cells with a hydrogen carrier,[17] but including geothermal energy where it is available; nuclear fission, if a fuel cycle could be devised that eliminates the economic, safety, and proliferation barriers which have so crippled its acceptance to date; and nuclear fusion, should its feasibility ever be demonstrated.

Global warming, climatic change, and energy policy are tightly intertwined. Much of contemporary energy policy is concerned with the identification and development of primary energy sources to replace oil and natural gas as these fuels are depleted. The only primary energy sources from which to choose are coal (and the other solid hydrocarbons), nuclear fission, and solar power. All other potential primary energy sources are too small to have other than local significance (geothermal and tidal power) or have yet to be proved technically feasible (controlled nuclear fusion). Solar power includes all of the many ways to collect solar energy—hydropower, wind, biomass, photovoltaics, direct heating, waves.

The stabilization of the atmospheric CO_2 level requires that most of the world's vast coal deposits not be used. Coal reserves, alone among the fossil fuels, are sufficiently great to produce at least a sixfold or eightfold increase in atmospheric carbon dioxide.

Reexamination of the nuclear option may be unavoidable. This reexamination should focus on the question: What criteria of proliferation resistance, waste management, passive safety, and cost control would nuclear power have to satisfy so that it would play a role in providing the world's energy in the twenty-first century? At the same time, there must be a continuing effort to avoid introducing new danger into the nuclear power system; in particular, the current plans to extend the commercialization of the reprocessing of plutonium, creating a new class of terrorist hazards and proliferation of nuclear-weapons states, should be reexamined.

The nuclear weapons proliferation risk which would accompany the replacement of fossil fuel with nuclear fission is staggering.[18] Global energy uses to sustain human activities now exceed 10 TW-years per year, 7 to 8 TW of which is derived from fossil fuels. Replacing the fossil fuel would require between 7,000 and 8,000 large nuclear power plants, each of which would produce about 1,000 kilograms of plutonium annually. Between 7 and 8 million kilograms of plutonium would therefore be produced per year and would be shipped hither and yon between the various facilities which constitute the nuclear power fuel cycle. The commercialization of plutonium would create major hazards because of its incredible toxicity and its potential use in weapons. Plutonium is the most toxic material ever considered for commerce: a millionth of a gram can cause lung cancer. A typical shipment would contain 250 kilograms—so there would be about 40,000 shipments annually. About 10 kilograms of plutonium is needed to construct an atomic bomb.

Increasing Energy End-Use Efficiency: Because of ongoing structural shifts to less energy-intensive activities in the economy and opportunities to make more efficient use of energy, it is technically and economically feasible to reduce U.S. per capita energy use by a factor of 2 while per capita GNP doubles.[19] The situation is similar in other industrial countries. The expansion of the delivery of goods and services in nonindustrialized societies can proceed with substantially lower per capita energy consumption than has historically been true in the industrialized countries.[20]

Improving energy end-use efficiency affords the least-cost option to reduce greenhouse gas emissions. These reductions can be realized in a fairly short time compared with the lead time needed to deploy new energy supply technology and will go a long way toward buying the time needed to develop a safe and economical energy supply that does not produce greenhouse gases (see Chapter 19).

LIMITING METHANE RELEASES

Methane production appears to derive primarily from agricultural activities rather than from industrialization (see Chapter 17). Various proposals have been advanced to reduce wetlands or rice paddies or to alter the internal metabolic processes of ruminants through contemporary bioscience so that they emanate something other than methane. None of these proposals appears promising.

A modification of diet would, however, have significant benefits in reducing global warming. The production of beef requires about ten times as much agricultural land as would the production of an equivalent amount of plant protein. Reducing beef production would reduce fossil fuel inputs into agriculture, reduce methane produced by the animals themselves, relieve at least some of the forces resulting in deforestation, and make more land available for reforestation.

Some methane releases result from venting from oil and gas fields, leaks from gas distribution systems, and the outgassing of coal mines. With a 50% reduction in fossil fuel use, methane emissions would decline by 5 to 10%.

REDUCING EMISSION OF THE CFCS

Specific industrial chemicals may be controlled—CFC-11 and CFC-12, for example, for which there has already been a call for a total ban on production. It has yet to be demonstrated, however, that compounds specified in the Montreal Protocol would not be replaced by compounds which would have less impact on stratospheric ozone but would nevertheless be greenhouse gases. As some of these compounds are important greenhouse gases, their release would have the potential for continued global warming.

REDUCING TROPOSPHERIC OZONE AND NITROUS OXIDE

Tropospheric ozone is produced as a result of the release of nitrogen oxides and carbon monoxide, largely produced by fossil fuel use. If carbon monoxide and NO_x emissions could be reduced, tropospheric ozone levels would decline. Nitrous oxide would also be reduced to the extent that fossil fuel combustion, especially that of coal, is reduced.

REDUCING GREENHOUSE GAS EMISSIONS AT LEAST COST TO SOCIETY

Greenhouse gas emissions can be reduced through several strategies:

- Reducing energy use by implementing known means to increase energy end-use efficiency

- Reforestation
- Replacing coal and oil in existing facilities by natural gas
- Deployment of biomass, photovoltaic, and other solar energy sources
- Shifting to diets with less animal protein and more plant protein
- Using CO_2 scrubbers on fossil fuel electricity generating plants
- Expansion of nuclear fission, as advocated by some

Implementing each of these strategies would require the commitment of capital and other resources. And the measures differ greatly in their cost effectiveness in reducing greenhouse gas emissions. Each dollar invested in increasing the efficiency of electricity use displaces several times as much CO_2 as does a dollar invested in nuclear power.[21] Improving auto and truck efficiency would be much more cost-effective in reducing CO_2 emissions than would replacing coal by natural gas in existing power plants.[22] Adding CO_2 scrubbers to coal-fired power plants would double the cost of electricity—a cost several times that which would be necessary to reduce greenhouse gas emissions by implementing known technologies to reduce energy consumption.[23] Preliminary analyses suggest that reforestation could reduce atmospheric CO_2 at low cost.[24]

CONCLUSIONS

Global warming considerations dramatically change energy and other policy options. Much of energy policy analysis has, in the past, been concerned with deciding what mix of primary energy sources can, and should, be deployed as oil and natural gas reserves are depleted. The implications of global warming are not usually covered in these analyses. It is assumed that there are three choices: coal and the other solid hydrocarbons; the renewables (solar, geothermal, and tidal power); and nuclear fission with breeder reactors. The issue is not yet resolved. During the 1960s the experts opted for fission, but when nuclear power began to move into the private sector and was exposed to the rigors of the market and public opinion, it was rejected in all but a few countries having strong central governmental control.

During the extensive deliberations regarding the acceptability of nuclear fission, it was assumed that both coal and the renewable energy sources were available, as it was recognized that coal's conventional environmental and health impacts could be controlled without abandoning the coal resource and at small incremental cost. It is now realized that to limit global warming requires that CO_2 emissions from fossil fuels must be reduced to less than 50% of present levels. This places coal in an entirely new light. There are few statements about the policy implications of global warming which can be made without reservation. One is that avoidance of global warming beyond 1 or 2°C over that to which we are already committed requires that coal use be reduced and that most of the coal and other solid hydrocarbon deposits remain in the ground. There can be no clean coal unless effective CO_2 scrubbers can be developed, and this seems unlikely.

Overall, reducing fossil fuel use is the single most important measure to control climatic change, as this results in reducing emissions of four major greenhouse gases: CO_2, methane, nitrous oxide, and tropospheric ozone. It appears that atmospheric CO_2 levels would stabilize if emissions were reduced by about 3 billion tons of carbon per year. This is possible.

Reducing the emissions of methane, CFCs, and other industrial greenhouse gases to levels not greater than their natural sinks (that is, to levels that will result in no net atmospheric buildup) appears, however, to be more difficult. To the extent that emissions of the non-CO_2 greenhouse gases cannot be reduced to their rate of removal from the atmosphere, CO_2 emissions must be further reduced.

In theory, it is possible to eventually halt continued global heating. If we act promptly—and if the most conservative of the scientific studies of climate sensitivity prove to be correct—greenhouse warming might be limited to about 2°C over that to which we are already committed. If we are slow to respond with policies to limit greenhouse gas releases— and the higher values of climate sensitivity are valid—then it may be impossible to eventually stabilize global temperature with less than an additional 3 to 4°C. In either case, however, eventual stabilization of global climate seems to require:

- A reduction in fossil fuel use by at least 50%. This would be achievable, first and foremost, through taking advantage of the vast potential for increasing the efficiency of energy use and to a lesser extent by the requirement that there be CO_2 removal from any new fossil-fueled central power station—by switching from oil and coal to natural gas in the short run and to solar and other renewable energy sources in the long run. Reducing fossil fuel use will also achieve reductions in the emission of several major greenhouse gases—CO_2, methane, nitrous oxide, and tropospheric ozone—and would also go a long way toward eliminating the causes of acid rain, urban smog, and other serious health problems and their economic costs.

- Shifting from deforestation to reforestation.
- A halt to the releases of CFC-11, CFC-12, and other long-lived industrial greenhouse gases and replacement of the ozone-depleting chemicals with substitutes which are not greenhouse gases.

Each of these measures is desirable for reasons other than stabilization of global climate. Each will be difficult to accomplish.

It must not be forgotten that global warming itself will increase atmospheric emissions of CO_2 and methane from biotic and marine carbon pools, but by how much is uncertain. It is necessary to limit the overall global warming so that large positive-feedback mechanisms are not engaged, as the potential releases from these sources are very large indeed.

Global climatic change is the central environmental problem of our time, and it is a situation in which there will be no winners. Each country will suffer the consequences of climatic change, and each must act to limit global warming to the fullest extent possible.

Notes

1. Board of Atmospheric Sciences and Climate, *Current Issues in Atmospheric Change* (Washington, D.C.: National Academy Press, 1987), 25–26.
2. It is usual to express the size of carbon movements and reservoirs in 10^9 tons of carbon (GT carbon). Some authors, however, express carbon dioxide fluxes in GT of carbon dioxide. One GT of carbon dioxide contains $12/44 = 0.27$ GT of carbon, or conversely 1 GT of carbon as carbon dioxide is the same as 3.67 GT of carbon dioxide.
3. There is both good ozone (in the stratosphere) and bad ozone (in the troposphere). Ozone in the upper atmosphere, the stratosphere, shields the earth from harmful levels of ionizing ultraviolet radiation from the sun. Were there no ozone shield, life on earth as we know it probably could not have developed. Stratospheric ozone is formed when incoming solar radiation interacts with ordinary oxygen, O_2. Ozone in the lower atmosphere, the troposphere, is, however, a noxious pollutant and a powerful greenhouse gas. Tropospheric ozone is formed largely by complex chemical reactions involving emissions from internal combustion engines. The chlorofluorocarbons, and other indus-

trial gases containing bromine, fluorine, and chlorine, destroy ozone in the stratosphere. In the troposphere, however, the CFCs are important greenhouse gases. Several articles listed at the back of the book discuss the atmospheric chemistry of the trace gases.

4. Detailed comparisons of carbon dioxide production from primary energy sources are found, for example, in G. J. MacDonald, *The Long-Term Impacts of Increasing Atmospheric Carbon Dioxide Levels* (Cambridge, Mass.: Ballinger, 1982).

5. Typical examples of such analyses can be found in the papers by L. Lave, R. D'Arge, and W. Nordhaus included in "The Global Commons: Costs and Climatic Effects," *American Economic Review 72* (1982). A somewhat more varied collection is included in *Journal of Policy Analysis and Management 7(3)*, where Lave largely repeats his arguments of 6 years earlier, P. G. Brown shows that welfare economics provides an unpromising framework for considering greenhouse warming policy, and I. Mintzer reviews technical responses which could at least delay global warming without threatening economic growth. Muddling through, another commonly espoused approach, is discussed by M. Glantz in "A Political View of CO_2," *Nature 280* (1979): 189–190.

6. Thomas C. Schelling, in Carbon Dioxide Assessment Committee, National Research Council, *Changing Climate* (Washington, D.C.: National Academy Press, 1983), 449–486.

7. An example: "The worst consequences [of greenhouse effect climate change] are likely to occur to unmanaged ecosystems and to developing countries; in contrast, the industrial nations may experience only minor irritations, apart from the eventual rise in sea level"; L. B. Lave, "The Greenhouse Effect: What Government Actions Are Needed," *Journal of Policy Analysis and Management 7(3)* (1988): 460–470.

8. See note 1 above.

9. Stephen H. Schneider, "The Greenhouse Effect: What We Can or Should Do About It," paper presented at the First North American Conference on Preparing for Climate Change: A Cooperative Approach, Washington, D.C., 1987.

10. Presently the 7 to 8 TW (tera watt 10^{12}) of fossil fuel use produces about 6 GT of carbon, as carbon dioxide, per year. So removal of the carbon dioxide from each TW-year per year of fossil fuel consumption would require about 0.7 to 1.4 million square kilometers of new forest.

11. For a detailed review of reforestation, see Gregg Marland, *The Prospect of Solving the CO_2 Problem Through Global Reforestation,* DOE/NBB-0082, U.S. Dept. of Energy, Washington, D.C., February 1988.

12. See, for example, William Broad, "Scientists Dream Up Bold Remedies for Ailing Atmosphere," *New York Times,* 16 August 1988.

13. See, for example, G. M. Hildy, *Hearings Before the Committee on Energy and Natural Resources U.S. Senate*, 248 ff., S. Hrg. 100–461, U.S. Government Printing Office, Washington, D.C. (9–10 November 1987); G. J. MacDonald, same hearings, pp. 238–286; J. F. Decker, same hearings, pp. 275–276; M. Steinberg et al., *A Systems Study for the Removal, Recovery, and Disposal of Carbon Dioxide from Fossil Fuel Power Plants in the U.S.*, DOE/CH/00016-2, U.S. Dept. of Energy, Washington, D.C. (December 1984); and B. Louks, "CO_2 Production in Gasification-Combined-Cycle Plants," *EPRI Journal* (October/November 1987): 52–54.

14. W. G. Snyder and C. A. Depew, *Coproduction of Carbon Dioxide and Electricity*, final report of Fluor Technology, Inc., research project, EPRI AP-4827, Electric Power Research Institute, November 1986.

15. Gregg Marland, "Technical Fixes for Limiting the Increase of Atmospheric CO_2: A Review," in press. Manuscript available from G. Marland, Institute for Energy Analysis, Oak Ridge, Tennessee, 37831-0117.

16. See note 4 above.

17. Joan M. Ogden and Robert H. Williams, "Solar Hydrogen and the Revolution in Amorphous Silicon Solar Cell Technology," in press, World Resources Institute.

18. Robert H. Williams, paper presented at the Workshop on Global Climatic Change, Woods Hole Research Center, Woods Hole, Massachusetts, 14 September 1988.

19. Robert H. Williams, "A Low-Energy Future for the United States," *Energy 12* (*10/11*) (1987): 929–955.

20. Perhaps the best energy supply and demand analysis to date is J. Goldemberg, T. B. Johansson, A.K.N. Reddy, and R. H. Williams, *Energy for a Sustainable World* (New Delhi: Wiley Eastern, 1988). An earlier summary by the same authors is found in J. Goldemberg, T. B. Johansson, A.K.N. Reddy, and R. H. Williams, "Basic Needs and Much More with One Kilowatt Per Capita," *Ambio 14* (*4–5*) (1985): 190. For an excellent and comprehensive exposition of energy conservation potentials, see *Energy 12* (*10/11*) (1987): Proceedings of the Soviet-American Symposium on Energy Conservation, Moscow, June 1985, R. H. Socolow and M. Ross, editors. See also William U. Chandler, *Energy Productivity: Key to Environmental Protection and Economic Progress*, Paper 63 (Washington, D.C.: Worldwatch Institute, 1985); J. Goldemberg, T. B. Johansson, A.K.N. Reddy, and R. H. Williams, "An End-Use Oriented Global Energy Strategy," *Annual Review of Energy 10* (1985); David J. Rose, M. M. Miller, and C. Agnew, *Global Energy Futures and CO_2-Induced Climate Change*, Report MITEL 83-015 (Cambridge, Mass.: MIT, 1984).

21. Bill Keepin and Gregory Kats, "Greenhouse Warming: Comparative Analysis of Two Abatement Strategies," *Energy Policy* (December 1988). See also

Charles Komanoff, "Greenhouse Effect Amelioration—Efficiency vs. Nuclear," memorandum dated 14 August 1988; available from KEA, 270 Lafayette St., Suite 902, New York, NY 10012.

22. William U. Chandler, personal communication.

23. William U. Chandler, Howard S. Geller, and Marc R. Ledbetter, *Energy Efficiency: A New Agenda* (Washington, D.C.: American Council for an Energy Efficient Economy, 1988); see notes 13 and 14 for the cost of CO_2 scrubbers.

24. Daniel Dudek, senior economist, Environmental Defense Fund, testimony before the U.S. Senate Committee on Energy and Natural Resources, 19 September 1988.

Chapter 2

THE GREENHOUSE EFFECT: IMPACTS ON CURRENT GLOBAL TEMPERATURE AND REGIONAL HEAT WAVES

James E. Hansen

- The earth is warmer in 1988 than at any time in the history of instrumental measurements.
- The global warming is now sufficiently large that we can ascribe with a high degree of confidence a cause and effect relationship to the greenhouse effect.

This statement was presented to the U.S. Senate Committee on Energy and Natural Resources on 23 June 1988. It is based largely on recent studies carried out by S. Lebedeff, D. Rind, I. Fung, A. Lacis, R. Ruedy, G. Russell, and P. Stone at the NASA Goddard Institute for Space Studies.

- In our computer climate simulations the greenhouse effect now is already large enough to begin to affect the probability of occurrence of extreme events such as summer heat waves. The model results imply that heat wave/drought occurrences in the Southeast and Midwest United States may be more frequent in the next decade than in climatological (1950–1980) statistics.

CURRENT GLOBAL TEMPERATURES

Present global temperatures are the highest in the period of instrumental records, as shown in Figure 2.1. The rate of global warming in the past two decades is higher than at any earlier time in the record. The four warmest years in the past century all have occurred in the 1980s.

The global temperature in 1988 up to June 1 is substantially warmer than the like period in any previous year in the record. This is illustrated in Figure 2.2, which shows seasonal temperature anomalies for the past few decades. The most recent two seasons (December–January–February and March–April–May 1988) are the warmest in the entire record. The first five months of 1988 are so warm globally that we conclude that 1988 will be the warmest year on record unless there is a remarkable, improbable cooling in the remainder of the year.

RELATIONSHIP OF GLOBAL WARMING AND GREENHOUSE EFFECT

Causal association of current global warming with the greenhouse effect requires determination that the warming is larger than natural climate variability, and the magnitude and nature of the warming is consistent with the greenhouse warming mechanism. Both of these issues are addressed quantitatively in Figure 2.3, which compares recent observed global temperature change with climate model simulations of temperature changes expected to result from the greenhouse effect.

The present observed global warming is close to 0.4°C, relative to "climatology," which is defined as the 30-year (1951–1980) mean. A

FIGURE 2.1
Global Temperature Trend: 1880–1988

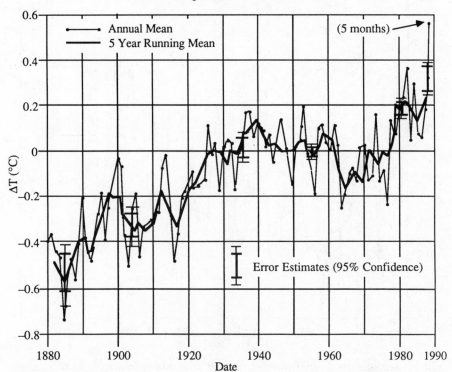

Global surface air temperature change for the past century, with the zero point defined as the 1951–1980 mean. Uncertainty bars (95% confidence limits) are based on an error analysis as described in reference 6; inner bars refer to the 5-year mean and outer bars to the annual mean. The analyzed uncertainty is a result of incomplete spatial coverage by measurement stations, primarily in ocean areas. The 1988 point compares the January–May 1988 temperature to the mean for the same 5 months in 1951–1980.

warming of 0.4°C is three times larger than the standard deviation of annual mean temperatures in the 30-year climatology. The standard deviation of 0.13°C is a typical amount by which the global temperature fluctuates annually about its 30-year mean; the probability of a chance warming of three standard deviations is about 1%. Thus we can state with about 99% confidence that current temperatures represent a real

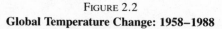

FIGURE 2.2

Global Temperature Change: 1958–1988

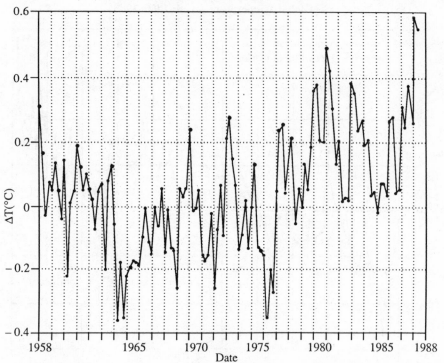

Global surface air temperature change at seasonal resolution for the past 30 years. Figures 2.1 and 2.2 are updates of results in reference 6.

warming trend rather than a chance fluctuation over the 30-year period.

We have made computer simulations of the greenhouse effect for the period since 1958, when atmospheric CO_2 began to be measured accurately. A range of trace gas scenarios is considered so as to account for moderate uncertainties in trace gas histories and larger uncertainties in future trace gas growth rates. The nature of the numerical climate model used for these simulations is described in reference 1. There are major uncertainties in the model, which arise especially from assumptions about (1) global climate sensitivity and (2) heat uptake and transport by the ocean, as discussed in reference 1. However, the magnitude of temperature changes computed with our climate model in various test

FIGURE 2.3
Computed Global Temperature Change: 1960–2020

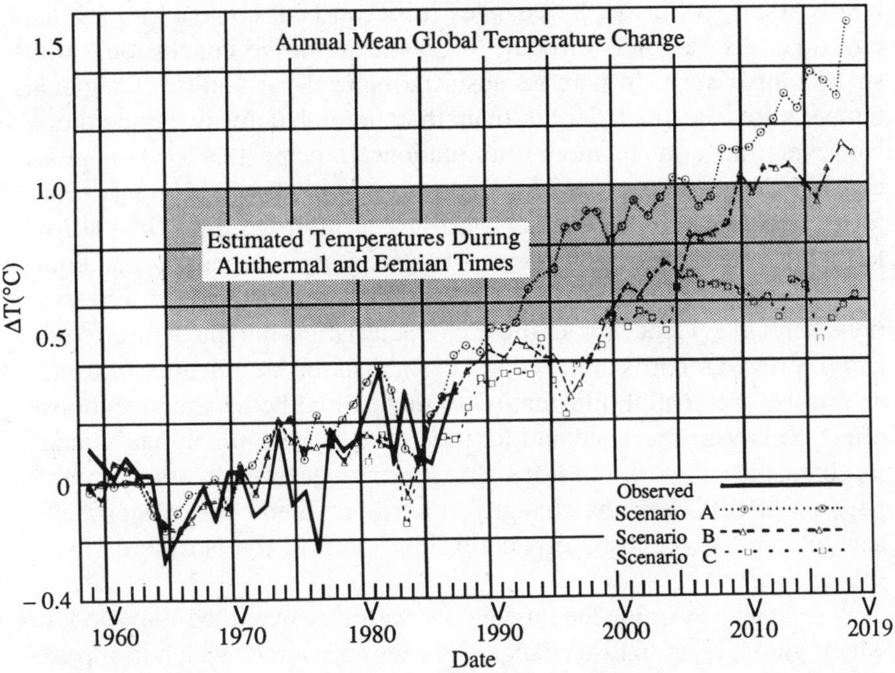

Annual mean global surface air temperature computed for trace gas scenarios A, B, and C described in reference 1. (Scenario A assumes continued growth rates of trace gas emissions typical of the past 20 years, i.e., about 1.5% year⁻¹ emission growth; scenario B has emission rates approximately fixed at current rates; scenario C drastically reduces trace gas emissions between 1990 and 2000.) Observed temperatures are from reference 6. The shaded range is an estimate of global temperature during the peak of the current and previous interglacial periods, about 6,000 and 120,000 years B.P., respectively. The zero point for observations is the 1951–1980 mean (reference 6); the zero point for the model is the control run mean.

cases is generally consistent with a body of empirical evidence (reference 2) and with sensitivities of other climate models (reference 1).

The global temperature change simulated by the model yields a warming over the past 30 years similar in magnitude to the observed warming (Figure 2.3). In both the observations and model the warming is close to 0.4°C by 1987, which is the 99% confidence level.

It is important to compare the spatial distribution of observed temperature changes with computer model simulations of the greenhouse effect, and also to search for other global changes related to the greenhouse effect, for example, changes in ocean heat content and sea ice coverage. As yet, it is difficult to obtain definitive conclusions from such comparisons, in part because the natural variability of regional temperatures is much larger than that of global mean temperature. However, the climate model simulations indicate that certain gross characteristics of the greenhouse warming should begin to appear soon, for example, somewhat greater warming at high latitudes than at low latitudes, greater warming over continents than over oceans, and cooling in the stratosphere while the troposphere warms. Indeed, observations contain evidence for all these characteristics, but much more study and improved records are needed to establish the significance of trends and to use the spatial information to understand better the greenhouse effect. Analyses must account for the fact that there are climate change mechanisms at work, besides the greenhouse effect; other anthropogenic effects, such as changes in surface albedo and tropospheric aerosols, are likely to be especially important in the Northern Hemisphere.

We can also examine the greenhouse warming over the full period for which global temperature change has been measured, which is approximately the past 100 years. On such a longer period the natural variability of global temperature is larger; the standard deviation of global temperature for the past century is 0.2°C. The observed warming over the past century is about 0.6–0.7°C. Simulated greenhouse warming for the past century is in the range 0.5–1.0°C, depending upon various modeling assumptions (e.g., reference 2). Thus, although there are greater uncertainties about climate forcings in the past century than in the past 30 years, the observed and simulated greenhouse warmings are consistent on both of these time scales.

CONCLUSION

Global warming has reached a level such that we can ascribe with a high degree of confidence a cause and effect relationship between the greenhouse effect and the observed warming. Certainly further study of this issue must be made. The detection of a global greenhouse signal represents only a first step in analysis of the phenomenon.

GREENHOUSE IMPACTS ON SUMMER HEAT WAVES

Global climate models are not yet sufficiently realistic to provide reliable predictions of the impact of greenhouse warming on detailed regional climate patterns. However, it is useful to make initial studies with state-of-the-art climate models; the results can be examined to see whether there are regional climate change predictions which can be related to plausible physical mechanisms. At the very least, such studies help focus the work needed to develop improved climate models and to analyze observed climate change.

One predicted regional climate change which has emerged in such climate model studies of the greenhouse effect is a tendency for mid-latitude continental drying in the summer (references 3, 4, 5). . . . Most of these studies have been for the case of doubled atmospheric CO_2, a condition which may occur by the middle of next century.

Our studies during the past several years at the Goddard Institute for Space Studies have focused on the expected transient climate change during the next few decades, as described in the attachment to my testimony. Typical results from our simulation for trace gas scenario B . . . show computed July temperature anomalies in several years between 1986 and 2029. In the 1980s the global warming is small compared to the natural variability of local monthly mean temperatures; thus the area with cool temperatures in a given July is almost as great as the area with warm temperatures. However, within about a decade the area with above normal temperatures becomes much larger than the area with cooler temperatures.

The specific temperature patterns for any given month and year should not be viewed as predictions for that specific time, because they depend upon unpredictable weather fluctuations. However, characteristics which tend to repeat warrant further study, especially if they occur for different trace gas scenarios. We find a tendency in our simulations of the late 1980s and the 1990s for greater than average warming in the Southeast and Midwest United States. . . . These areas of high temperature are usually accompanied by below normal precipitation.

Examination of the changes in sea level pressure and atmospheric winds in the model suggests that the tendency for larger than normal warming in the Midwest and Southeast is related to the ocean's response time; the relatively slow warming of surface waters in the mid-Atlantic off the eastern United States and in the Pacific off California tends to increase sea level pressure in those ocean regions and this in turn tends to cause more southerly winds in the eastern United States and more northerly winds in the western United States. However, the tendency is too small to be apparent every year; in some years in the 1990s the eastern United States is cooler than climatology (the control run mean).

CONCLUSION

It is not possible to blame a specific heat wave/drought on the greenhouse effect. However, there is evidence that the greenhouse effect increases the likelihood of such events; our climate model simulations for the late 1980s and the 1990s indicate a tendency for an increase of heat wave/drought situations in the Southeast and Midwest United States. We note that the correlations between climate models and observed temperatures are often very poor at subcontinental scales, particularly during Northern Hemisphere summer (reference 7). Thus improved understanding of these phenomena depends upon the development of increasingly realistic global climate models and upon the availability of global observations needed to verify and improve the models.

References

1. Hansen, J., I. Fung, A. Lacis, D. Rind, G. Russell, S. Lebedeff, R. Ruedy, and P. Stone, 1988, Global climate changes as forecast by the GISS 3-D model, *J. Geophys. Res.* (in press).
2. Hansen, J., A. Lacis, D. Rind, G. Russell, P. Stone, I. Fung, R. Ruedy, and J. Lerner, 1984, Climate sensitivity: analysis of feedback mechanisms, *Geophys. Mono., 29,* 130–163.
3. Manabe, S., R. T. Wetherald, and R. J. Stauffer, 1981, Summer dryness due to an increase in atmospheric CO_2 concentration, *Climate Change, 3,* 347–386.
4. Manabe, S., and R. T. Wetherald, 1986, Reduction in summer soil wetness induced by an increase in atmospheric carbon dioxide, *Science, 232,* 626–628.
5. Manabe, S., and R. T. Wetherald, 1987, Large-scale changes of soil wetness

induced by an increase in atmospheric carbon dioxide, *J. Atmos. Sci., 44,* 1211–1235.

6. Hansen, J., and S. Lebedeff, 1987, Global trends of measured surface air temperature, *J. Geophys. Res., 92,* 13345–13372; Hansen, J., and S. Lebedeff, 1988, Global surface air temperatures: update through 1987, *Geophys. Res. Lett., 15,* 323–326.

7. Grotch, S., 1988, Regional intercomparisons of general circulation model predictions and historical climate data, Dept. of Energy Report, DOE/NBA-0084.

Chapter 3

THE CHANGING ATMOSPHERE: IMPLICATIONS FOR GLOBAL SECURITY

Toronto Conference

FOREWORD
H. L. Ferguson
Conference Director

At the invitation of the Government of Canada, more than 300 world experts—leaders in science, law, and the environment; ministers of government; economists; industrialists; policy analysts; and officials from international agencies—assembled in Toronto, Ontario, Canada, from June 27–30, 1988, to consider the threats posed by the changing global atmosphere and how they might be addressed. They came from 46 countries and quickly arrived at a consensus that the concerns about the effects of atmospheric change—greenhouse gases, ozone-layer de-

Conference statement, Toronto, Ontario, Canada, 27–30 June 1988.

pleting substances, toxics, smog, and acid rain—are justified and that
the time to act on the problems is now. The Conference was the first
direct response to the call for action of the UN's World Commission on
Environment and Development. It was also the first comprehensive
meeting between specialists on the issues at hand and high-level policymakers. The significance of the event was underscored by the participation of Prime Ministers Mulroney of Canada and Brundtland of Norway,
the participation of Ministers McMillan and Masse (Canada), Salim
(Indonesia), Nijpels (Netherlands), Cissokho (Senegal), Luttenbarck
Batalha (Brazil), Harilla (Morocco), by Senator Wirth (United States),
and by ambassadors from Algeria, Canada, The Maldives, and
Sweden.

The message from the Toronto Conference was clear. The earth's
atmosphere is being changed at an unprecedented rate, primarily by
humanity's ever-expanding energy consumption, and these changes
represent a major threat to global health and security. Sound policies
must be quickly developed and implemented to provide for the protection of the planet's atmosphere. That message and an agenda for action
are embodied in this statement of the Conference's conclusions and
recommendations. The statement builds on important preceding conferences and workshops, and draws heavily from ideas and discussion of
the Conference's 12 Working Groups. Its careful reading is recommended to all decision-makers seeking solutions to the problems of
climate change.

I wish to take this opportunity to thank my colleagues on the Conference Statement Committee. These colleagues, who worked long and
difficult hours in drafting the Conference Statement and who also
served as advisors on Conference planning over the past two years,
are J. P. Bruce, G. Goodman, J. Jaeger, G. A. McKay, J. MacNeill,
M. Oppenheimer, and P. Usher. Dr. Jaeger also produced the main
background paper to the Conference. In addition I must thank the
Conference General Chairman, Canada's Ambassador to the United
Nations, Stephen Lewis, for his important contributions to the final draft
of the statement.

My thanks also go to the many international experts who wrote the
theme papers that provided background to the Conference discussions,
to the chairpersons and rapporteurs who so skillfully managed the
Working Group sessions, to those who assumed special speaking as-

signments, and to persons and groups who prepared special reports for Working Group discussions and for general consideration by the Conference. Finally, I extend my deep gratitude to all who participated in the Conference—delegates, observers, media, and staff—and thereby contributed to its outstanding success. Their collective efforts constitute a landmark in confronting one of humankind's biggest challenges.

I believe the Conference will prove to have been an important step forward in reconciling environmental, societal, and developmental goals. We still have a long way to go. However, I am confident that the Toronto Conference gave us the right agenda and conviction to act. It also provided an opportunity to share our views with world leaders from many disciplines—scientific, social, and political.

SUMMARY

Humanity is conducting an unintended, uncontrolled, globally pervasive experiment whose ultimate consequences could be second only to a global nuclear war. The earth's atmosphere is being changed at an unprecedented rate by pollutants resulting from human activities, inefficient and wasteful fossil fuel use, and the effects of rapid population growth in many regions. These changes represent a major threat to international security and are already having harmful consequences over many parts of the globe.

Far-reaching impacts will be caused by global warming and sea-level rise, which are becoming increasingly evident as a result of the continued growth in atmospheric concentrations of carbon dioxide and other greenhouse gases. Other major impacts are occurring from ozone-layer depletion resulting in increased damage from ultraviolet radiation. The best predictions available indicate potentially severe economic and social dislocation for present and future generations, which will worsen international tensions and increase risk of conflicts between and within nations. It is imperative to act now.

These were the major conclusions of the World Conference on the Changing Atmosphere: Implications for Global Security, held in Toronto, Ontario, Canada, June 27–30, 1988. More than 300 scientists and policymakers from 46 countries, United Nations organizations, other international bodies, and nongovernmental organizations participated in the sessions.

The Conference called upon governments, the United Nations and its specialized agencies, industry, educational institutions, nongovernmental organizations, and individuals to take specific actions to reduce the impending crisis caused by pollution of the atmosphere. No country can tackle this problem in isolation. International cooperation in the management and monitoring of, and research on, this shared resource is essential.

The Conference called upon governments to work urgently towards an *Action Plan for the Protection of the Atmosphere*. This should include an international framework convention, while encouraging other standard-setting agreements along the way, as well as national legislation to provide for protection of the global atmosphere. The Conference also called upon governments to establish a *World Atmosphere Fund* financed in part by a levy on the fossil fuel consumption of industrialized countries to mobilize a substantial part of the resources needed for these measures.

THE ISSUE

Continuing alteration of the global atmosphere threatens global security, the world economy, and the natural environment through:

- Climate warming, rising sea level, altered precipitation patterns, and changed frequencies of climatic extremes induced by the "heat trap" effects of greenhouse gases.
- Depletion of the ozone layer.
- Long-range transport of toxic chemicals and acidifying substances.

These changes will:

- Imperil human health and well-being.
- Diminish global food security, through increases in soil erosion and greater shifts and uncertainties in agricultural production, particularly for many vulnerable regions.
- Change the distribution and seasonal availability of freshwater resources.
- Increase political instability and the potential for international conflict.
- Jeopardize prospects for sustainable development and the reduction of poverty.
- Accelerate the extinction of animal and plant species upon which human survival depends.

- Alter yield, productivity, and biological diversity of natural and managed ecosystems, particularly forests.

If rapid action is not taken now by the countries of the world, these problems will become progressively more serious, more difficult to reverse, and more costly to address.

SCIENTIFIC BASIS FOR CONCERN

The Conference calls for urgent work on an *Action Plan for the Protection of the Atmosphere.* This Action Plan, complemented by national action, should address the problems of climate warming, ozone layer depletion, long-range transport of toxic chemicals, and acidification.

CLIMATE WARMING

1. There has been an observed increase of globally averaged temperature of 0.7°C in the past century which is consistent with theoretical greenhouse gas predictions. The accelerating increase in concentrations of greenhouse gases in the atmosphere, if continued, will probably result in a rise in the mean surface temperature of the earth of 1.5 to 4.5°C before the middle of the next century.
2. Marked regional variations in the amount of warming are expected. For example, at high latitudes the warming may be twice the global average. Also, the warming would be accompanied by changes in the amount and distribution of rainfall and in atmospheric and ocean circulation patterns. The natural variability of the atmosphere and climate will continue and be superimposed on the long-term trend, forced by human activities.
3. If current trends continue, the rates and magnitude of climate change in the next century may substantially exceed those experienced over the last 5,000 years. Such high rates of change would be sufficiently disruptive that no country would likely benefit *in toto* from climate change.
4. The climate change will continue so long as the greenhouse gases accumulate in the atmosphere.
5. There can be a time lag of the order of decades between the emission of gases into the atmosphere and their full manifestation in atmospheric and biological consequences. Past emissions have already committed planet Earth to a significant warming.

6. Global warming will accelerate the present sea-level rise. This will probably be of the order of 30 cm but could possibly be as much as 1.5 m by the middle of the next century. This could inundate low-lying coastal lands and islands, and reduce coastal water supplies by increased saltwater intrusion. Many densely populated deltas and adjacent agricultural lands would be threatened. The frequency of tropical cyclones may increase and storm tracks may change with consequent devastating impacts on coastal areas and islands by floods and storm surges.

7. Deforestation and bad agricultural practices are contributing to desertification and are reducing the biological storage of carbon dioxide, thereby contributing to the increase of this most important greenhouse gas. Deforestation and poor agricultural practices are also contributing additional greenhouse gases such as nitrous oxide and methane.

OZONE LAYER DEPLETION

1. Increased levels of damaging ultraviolet radiation, while the stratospheric ozone shield thins, will cause a significant rise in the occurrence of skin cancer and eye damage and will be harmful to many biological species. Each 1% decline in ozone is expected to cause a 4 to 6% increase in certain kinds of skin cancer. A particular concern is the possible combined effects on unmanaged ecosystems of both increased ultraviolet radiation and climate changes.

2. Over the last decade, a decline of 3% in the ozone layer has occurred at mid-latitudes in the Southern Hemisphere, possibly accompanying the appearance of the Antarctic ozone hole; although there is more meteorological variability, there are indications that a smaller decline has occurred in the Northern Hemisphere. Changes of the ozone layer will also change the climate and the circulation of the atmosphere.

ACIDIFICATION

In improving the quality of the air in their cities, many industrialized countries unintentionally sent increasing amounts of pollution across national boundaries in Europe and North America, contributing to the acidification of distant environments. This was manifested by increasing damage to lakes, soils, plants, animals, forests, and fisheries. Failure to control automobile pollution in some regions has seriously contributed to the problem. The principal damage agents are oxides of sulfur and nitrogen as well as volatile hydrocarbons. The resulting acids can also corrode buildings and metallic structures causing overall billions of dollars of damage annually.

The various issues arising from the pollution of Earth's atmosphere by a number of substances are often closely interrelated, both through chemistry and through potential control strategies. For example, chlorofluorocarbons (CFCs) both destroy ozone and are greenhouse gases; conservation of fossil fuels would contribute to addressing both acid rain and climate change problems.

SECURITY: ECONOMIC AND SOCIAL CONCERNS

As the *UN Report on the Relationship Between Disarmament and Development* states: "The world can either continue to pursue the arms race with characteristic vigor or move consciously and with deliberate speed toward a more stable and balanced social and economic development within a more sustainable international economic and political order. It cannot do both. It must be acknowledged that the arms race and development are in a competitive relationship, particularly in terms of resources, but also in the vital dimension of attitudes and perceptions." The same consideration applies to the vital issue of protecting the global atmospheric commons from the growing peril of climate change and other atmospheric changes. Unanticipated and unplanned change may well become the major nonmilitary threat to international security and the future of the global economy.

There is no concern more fundamental than access to food and water. Currently, levels of global food security are inadequate but even those will be most difficult to maintain into the future, given projected agricultural production levels and population and income growth rates. The climate changes envisaged will aggravate the problem of uncertainty in food security. Climate change is being induced by the prosperous, but its effects are suffered most acutely by the poor. It is imperative for governments and the international community to sustain the agricultural and marine resource base and provide development opportunities for the poor in light of this growing environmental threat to global food security.

The countries of the industrially developed world are the main source of greenhouse gases and therefore bear the main responsibility to the world community for ensuring that measures are implemented to address the issues posed by climate change. At the same time, they must see that the developing nations of the world, whose problems are greatly aggravated by population growth, are assisted in and not inhibited from improving their economies and the living conditions of their citizens. This will necessitate a wide range of measures, including significant additional energy use in those countries and compensating reductions in the industrialized countries. The transition to a sustainable future will require investments in energy efficiency and nonfossil energy sources. In order to ensure that these investments occur, the global community must not only halt the current net transfer of resources from developing countries, but actually reverse it. This reversal should embrace the technologies involved, taking into account the implications for industry.

A coalition of reason is required, in particular, a rapid reduction of both North–South inequalities and East–West tensions, if we are to achieve the understanding and agreements needed to secure a sustainable future for planet Earth and its inhabitants.

It takes a long time to develop an international consensus on complex issues such as these, to negotiate, sign, and ratify international environmental instruments, and to begin to implement them. It is therefore imperative that serious negotiations start now.

LEGAL ASPECTS

The first steps in developing international law and practices to address pollution of the air have already been taken: in the Trail Smelter arbitration of 1935 and 1938; Principle 21 of the 1972 Declaration of the UN Conference on the Environment; the Economic Commission for Europe (ECE) Convention on Long Range Transboundary Air Pollution and its protocol (Helsinki, 1985) for sulfur reductions, Part XII of the Law of the Sea Convention: and the Vienna Convention for Protection of the Ozone Layer and its Montreal Protocol (1987).

These are important first steps and should be actively implemented and respected by all nations. However, there is no overall convention constituting a comprehensive international framework that can address the interrelated problems of the global atmosphere or that is directed towards the issues of climate change.

A CALL FOR ACTION

The Conference urges immediate action by governments, the United Nations and their specialized agencies, other international bodies, non-governmental organizations, industry, educational institutions, and individuals to counter the ongoing degradation of the atmosphere.

An *Action Plan for the Protection of the Atmosphere* needs to be developed, which includes an international framework convention that encourages other standard-setting agreements and national legislation to provide for the protection of the global atmosphere. This must be complemented by implementation of national action plans that address the problems posed by atmospheric change (climate warming, ozone layer depletion, acidification, and the long-range transport of toxic chemicals) at their roots.

The following actions are mostly designed to slow and eventually reverse deterioration of the atmosphere. There are also a number of strategies for adapting to changes that must be considered. These are dealt with primarily in the recommendations of the Working Groups.

ACTIONS BY GOVERNMENTS AND INDUSTRY

- Ratify the Montreal Protocol on Substances That Deplete the Ozone Layer. The protocol should be revised in 1990 to ensure nearly complete elimination of the emissions of fully halogenated CFCs by the year 2000. Additional measures to limit other ozone-destroying halocarbons should be considered.
- Set energy policies to reduce the emissions of CO_2 and other trace gases in order to reduce the risks of future global warming. Stabilizing the atmospheric concentrations of CO_2 is an imperative goal. It is currently estimated to require reductions of more than 50% from present emission levels. Energy research and development budgets must be massively directed to energy options which would eliminate or greatly reduce CO_2 emissions and to studies undertaken to further refine the target reductions.

- Reduce CO_2 emissions by approximately 20% of 1988 levels by the year 2005 as an initial global goal. Clearly, the industrialized nations have a responsibility to lead the way, both through their national energy policies and their bilateral and multilateral assistance arrangements. About one-half of this reduction would be sought from energy efficiency and other conservation measures. The other half should be effected by modifications in supplies.
- Set targets for energy efficiency improvements that are directly related to reductions in CO_2 and other greenhouse gases. A challenging target would be to achieve the 10% energy efficiency improvements by 2005. Improving energy efficiency is not precisely the same as reducing total carbon emissions and the detailed policies will not all be familiar ones. A detailed study of the systems implications of this target should be made. Equally, targets for energy supply should also be directly related to reductions in CO_2 and other greenhouse gases. As with efficiency, a challenging target would again be to achieve the 10% energy supply improvements by 2005. A detailed study of the systems implications of this target should also be made. The contributions to achieving this goal will vary from region to region; some countries have already demonstrated a capability for increasing efficiency by more than 2% a year for over a decade.

   Apart from efficiency measures, the desired reduction will require (1) switching to lower CO_2 emitting fuels, (2) reviewing strategies for the implementation of renewable energy, especially advanced biomass conversion technologies; (3) revisiting the nuclear power option, which lost credibility because of problems related to nuclear safety, radioactive wastes, and nuclear weapons proliferation. If these problems can be solved, through improved engineering designs and institutional arrangements, nuclear power could have a role to play in lowering CO_2 emissions.
- Negotiate now on ways to achieve the above-mentioned reductions.
- Initiate management systems in order to encourage, review, and approve major new projects for energy efficiency.
- Vigorously apply existing technologies, in addition to gains made through reduction of fossil fuel combustion, to reduce (1) emissions of acidifying substances to reach the critical load that the environment can bear; (2) substances which are precursors of tropospheric ozone; and (3) other non-CO_2 greenhouse gases.
- Label products to allow consumers to judge the extent and nature of the atmospheric contamination that arises from the manufacture and use of the product.

ACTIONS BY MEMBER GOVERNMENTS OF THE UNITED
NATIONS, NONGOVERNMENTAL ORGANIZATIONS,
AND RELEVANT INTERNATIONAL BODIES

- Initiate the development of a comprehensive global convention as a framework for protocols on the protection of the atmosphere. The convention should emphasize such key elements as the free international exchange of information and the support of research and monitoring, and should provide a framework for specific protocols for addressing particular issues, taking into account existing international law. This should be vigorously pursued at the International Workshop on Law and Policy to be held in Ottawa early in 1989, the high-level political conference on Climate Change in the Netherlands in the Fall, 1989, the World Energy Conference in Canada in 1989 and the Second World Climate Conference in Geneva, June 1990, with a view to having the principles and components of such a convention ready for consideration at the Intergovernmental Conference on Sustainable Development in 1992. These activities should in no way impede simultaneous national, bilateral, and regional actions and agreements to deal with specific problems such as acidification and greenhouse gas emissions.
- Establish a World Atmosphere Fund, financed in part by a levy on fossil fuel consumption of industrialized countries, to mobilize a substantial part of the resources needed for implementation of the *Action Plan for the Protection of the Atmosphere.*
- Support the work of the Intergovernmental Panel on Climate Change to conduct continuing assessments of scientific results and to initiate government-to-government discussion of responses and strategies.
- Devote increasing resources to research and monitoring efforts within the World Climate Program, the International Geosphere Biosphere Program, and Human Response to Global Change Program. It is particularly important to understand how climate changes on a regional scale are related to an overall global change of climate. Emphasis should also be placed on better determination of the role of oceans in global heat transport and the flux of greenhouse gases.
- Increase significantly the funding for research, development, and transfer of information on renewable energy, if necessary by the establishment of additional and bridging programs; extend technology transfer with particular emphasis on the needs of the developing countries; and upgrade efforts to meet obligations for the development and transfer of technology embodied in existing agreements.

- Expand funding for more extensive technology transfer and technical cooperation projects in coastal zone protection and management.
- Reduce deforestation and increase afforestation making use of proposals such as those in the World Commission on Environment and Development's (WCED) report, "Our Common Future," including the establishment of a trust fund to provide adequate incentives to enable developing nations to manage their tropical forest resources sustainably.
- Develop and support technical cooperation projects to allow developing nations to participate in international mitigation efforts, monitoring, research, and analysis related to the changing atmosphere.
- Ensure that this Conference Statement, the Working Group reports, and the full Proceedings of the World Conference, "The Changing Atmosphere: Implications for Global Security" . . . are made available to all nations, to the conferences mentioned above, and to other future meetings dealing with related issues.
- Increase funding to nongovernmental organizations to allow the establishment and improvement of environmental education programs and public awareness campaigns related to the changing atmosphere. Such programs would aim at sharpening perception of the issues, and changing public values and behavior with respect to the environment.
- Allocate financial support for environmental education in primary and secondary schools and universities. Consideration should be given to establishing special groups in university departments for addressing the crucial issues of global climate change.

SPECIFIC RECOMMENDATIONS
OF WORKING GROUPS

The recommended actions in the Conference Statement are mostly general in nature and common to a number of Conference Working Groups. The specific recommendations of the Working Groups are given in the following section.

ENERGY

1. Targets for energy supply should be directly related to reductions in CO_2 and other greenhouse gases. A challenging target would be to reduce the annual global CO_2 emissions by 20% by the year 2005

through improved energy efficiency, altered energy supply, and energy conservation.

2. Research and demonstration projects should be undertaken to accelerate the development of advanced biomass conversion technologies.
3. Deforestation should be reduced and reforestation accelerated to significantly reduce the atmospheric concentrations of CO_2 and to replenish the primary fuel supply for the majority of the world's population.
4. There is a need to revisit the nuclear power option. If the problems of safety, waste, and nuclear arms proliferation can be solved, nuclear power could have a role to play in lowering CO_2 emissions.
5. It is necessary to internalize externalized costs. Policies should be fashioned to achieve broad, complementary social objectives and to minimize total social, economic, and environmental costs.

FOOD SECURITY

1. National governments are urged to reduce the contributions of agricultural activities to the concentration of greenhouse gases in the atmosphere. These contributions arise from the destruction of forests, the inefficient use of inorganic nitrogen fertilizers, the increased conversion of land to paddy rice cultivation, and the increased number of ruminant animals.
2. National governments should take the prospect of climate change into account in long-term agricultural and food security planning, particularly with respect to food availability to the most vulnerable groups.
3. National governments and international agencies should give increasing emphasis to a wide array of policy measures to reduce the sensitivity of the food supply to climatic variability in order to increase resilience and adaptability to climate change.
4. National governments are urged to increase their efforts to build subregional and regional cooperation aimed at achieving food security. International agencies should assist in promoting these regional cooperative efforts.
5. FAO, World Bank, WMO, UNEP, UNDP, CGIAR, and other international organizations should encourage research leading to ecologically sound agricultural management systems.

URBANIZATION AND SETTLEMENT

1. Environmental impact statements and land-use management plans should consider future climatic conditions including the local effects of rising sea level on coastal communities.

2. Urban authorities should undertake risk assessments and develop emergency planning procedures that take into account the effects of climate change, for example, the increased incidence of natural hazards.

3. National governments and the international aid community should develop policies and actions to deal with the likely increased movements of environmental refugees resulting from climate change.

4. Environmental education must be stressed, particularly with respect to the sustainable development of urban areas and human settlements, and should be strongly promoted by local and national authorities and by international bodies such as WMO, UNCHS, UNEP, UNIDO, and UNDP.

5. Comprehensive worldwide assessments should be made by national and international organizations of the vulnerability of specific geographic regions and urban areas to the increased risk of higher incidence and spread of infectious diseases due to global climate change, including both vector-borne and communicable diseases. In these areas, assessments should be made of health care infrastructures and of their ability to cope with the projected increased risks of the spread of infectious diseases; and steps should be identified to be taken by local and national authorities and international organizations to improve such capabilities.

6. Assessments should be made of the vulnerability of nuclear facilities, municipal and hazardous waste dumps, and other waste disposal facilities to the increased hazard of sudden flooding or gradual inundation, and of their potential for the consequent spread of infectious pathogens or toxic chemicals to the surrounding land and sea areas, and appropriate steps should be taken to minimize such risks.

WATER RESOURCES

1. The efficiency of water use and the resilience of existing and planned water resource systems and management processes must be increased to meet the existing climate variability.

2. Existing acid rain conventions must be extended to the global scale and modified to include toxic organic pollutants.

3. Integrated monitoring and research programs are urgently required to improve the methods of assessing the sensitivity of water resource systems, to identify critical regions and river basins where changes in hydrological processes and water demand will cause serious problems, and to understand and model the hydrological, ecological, and socioeconomic impacts of climate change.

4. To alleviate present and future water problems and to achieve sustainable development, we strongly endorse the global principle of interregional and intergenerational equity in all actions. International cooperation, open technology transfer, meaningful public involvement, and effective public information programs are essential.

LAND RESOURCES

An international fund should be created specifically for development assistance and research in order to:

1. Maintain the terrestrial reservoirs of carbon through the careful management and protection of tropical and temperate forests and their soils, tundra, and wetlands that represent major carbon pools.
2. Encourage the development of varieties of sustainable land-use practices through such activities as agroforestry, reforestation, development of varieties for adaptation to climate change, and development of effective management practices for waste treatment and disposal, and through policies for the use, settlement, and tenure of land. This requires major changes in the aid policy, commercial practices, and policies of related organizations (ITTO, FAO/TFAP, and ICRAF) as well as possible "debt swapping" for forest protection and access to a reforestation fund.
3. Identify the most productive agricultural lands so as to be able to implement a land reserve system that can be used to mitigate losses resulting from a more adverse climate and sea level rise.
4. Increase awareness among the public of issues posed by climate change in relation to the continued wise use of lands in a sustainable manner.
5. Broaden existing programs that address the impact on land resources of acid and other toxic depositions, by taking account of their global dimension.

COASTAL AND MARINE RESOURCES

1. Research is required to understand which natural and human factors determine the productivity and variability of marine and coastal resources.
2. Institutional and legal arrangements for the wise use of common property resources must be greatly improved.
3. The flexibility of marine-dependent industries and coastal commu-

nities must be greatly enhanced to respond to climate-induced changes.

4. Site-specific impact studies of the effects of sea-level rise must be undertaken. These should include consideration of the human, economic, and environmental risks and should result in local education programs.

5. The implications of climate change for coastal-zone planning must be considered, particularly the risk of sea-level rise and/or the potential need to locate new developments inland.

FUTURES AND FORECASTING

1. In order to have any hope of coping with future change, we must acquire and make use of the knowledge of the past and develop the ability to anticipate the possible future. No one model can or should be expected to deal with the uncertainties in forecasting, the details needed for making decisions, and the social, technical, and economic implications of change. Hence an array of techniques must be used in order to produce useful results.

2. Not only are continued efforts needed to improve forecasting methodologies and to integrate cause-and-effect modeling, but also improvements are needed in our ability to communicate and convey their implications for the broader culture so that individual and collective decisions can be made appropriately and with foresight. Attitudinal and institutional changes will be necessary because of the projected serious global consequences. Equally important is the need to take action, in an environmentally sustainable way, on the interrelated issues of population growth, resource use and depletion, and technological inequalities.

DECISION-MAKING AND UNCERTAINTY

1. The reduction of uncertainties requires advanced understanding of the chemistry of the atmosphere, of the implications of climate change for health, agriculture, economies, and other social concerns, and of the legal, political, and other aspects of the possible responses to climate change (prevention, compensation, and adaptation).

2. The industrialized nations should begin to restore the integrity of the environment, making atmospheric change the turning point for an ecological innovation of industrial economy.

3. Emission targets ought to be the subject of an international treaty between the nations that take the first step. Those nations should invite all the others to join them in advancing environmentally sustainable economic development.

4. Open decision-making may well provide for decisions that are not easily accepted by the public. We recommend a democratic discussion about possible responses to the atmospheric threat. Nongovernmental organizations should play a decisive role in furthering this discourse.

INDUSTRY, TRADE, AND INVESTMENT

Proposed as matters for urgent action are:

1. Creation of a World Atmosphere Fund financed by a levy on the fossil fuel consumption of industrialized countries, sufficient to support development and transfer fuel-efficient technologies.

2. Development of mechanisms for incorporating environmental considerations and responsibilities into the internal decision and reporting processes of business and industry.

3. Formation of an international consultative mechanism at the highest level, reporting to heads of government, to assure:

- Accelerated research and development efforts
- Reduction of institutional barriers to the adoption of appropriate low-emission technologies by industries and households
- Improvement of market information to promote the shift of consumption toward ecologically appropriate products.

GEOPOLITICAL ISSUES

1. The particular regions of the world or sectors of the economy that will be damaged first or most strongly by a rapidly changing atmosphere cannot be foreseen today, but the magnitude and variety of the eventual impacts are such that it is in the self-interest of all people to join in prompt action to slow the change and to negotiate toward an international accord on achieving shared responsibility for care of the climate and the atmosphere.

2. Coordinated international efforts and an all-encompassing international agreement are required along with prompt action by governmental agencies and nongovernmental groups to prevent harmful

changes to the atmosphere. Such actions can be based on improvements in energy efficiency, the use of alternative energy sources, and the transfer of technology and resources to the Third World.

LEGAL DIMENSIONS

1. More states should observe the international principles and norms that exist and all should be encouraged to enact or strengthen appropriate national legislation for the protection of the atmosphere.
2. The offer of the Prime Minister of Canada to host a meeting of law and policy experts in early 1989 should be accepted. That meeting should address the question of the progressive development and codification of the principles of international law taking into account the general principles of law set out in the Trail smelter, Lac Lanoux, Corfu Channel cases, Principle 21 of the 1972 Declaration of the United Nations Conference on the Human Environment, the Convention on Long-Range Transboundary Air Pollution and related protocols, Part XII of the Law of the Sea Convention and the Vienna Convention for the Protection of the Ozone Layer and its Montreal Protocol. The meeting should be directed toward the elaboration of the principles to be included in an umbrella/framework Convention on the Protection of the Atmosphere—one that would lend itself to the development of specific agreements/protocols laying down international standards for the protection of the atmosphere, in addition to existing instruments.

INTEGRATED PROGRAMS

1. A thorough review is required to establish the institutional needs for cooperation in research, impact assessment, and development of public policy options at the international, intergovernmental, and nongovernmental levels, at regional levels, and at national levels. This review should be completed by 1992.
2. Extension and further development are required for a United Nations global monitoring and information system that will incorporate technological advances in measurement, data storage and retrieval, and communications in order to track systematic changes in the physical, chemical, biological, and socioeconomic parameters that collectively describe the total global human environment. The responsibility for

development rests with governments. The monitoring system should be in place by the year 2000.

3. Also required is the development of an educational program to familiarize present and future generations with the importance of addressing issues concerning sustainable development including the actions and integrated, interdisciplinary programs needed.

Chapter 4

THE SCIENTIFIC CONSENSUS

Villach (Austria) Conference

CONFERENCE STATEMENT

A joint UNEP/WMO/ICSU Conference was convened in Villach, Austria, from 9 to 15 October 1985, with scientists from 29 developed and developing countries, to assess the role of increased carbon dioxide and other radiatively active constituents of the atmosphere (collectively known as greenhouse gases and aerosols) on climate changes and associated impacts. The other greenhouse gases reinforce and accelerate the impact due to CO_2 alone. As a result of the increasing concentrations of greenhouse gases, it is now believed that in the first half of the next century a rise of global mean temperature could occur which is greater than any in man's history.

Report of the International Conference on the Assessment of the Role of Carbon Dioxide and Other Greenhouse Gases in Climate Variations and Associated Impacts, Villach, Austria, 9–15 October 1985. The scientific papers prepared for the conference, and the conference statement, have been published as *The Greenhouse Effect, Climatic Change, and Ecosystems (SCOPE 29)* (Chichester: Wiley, 1986). Source: UNEP/WMO/ICSU Villach Conference Statement, WMO-NO. 661. Reprinted with permission by the World Meteorological Organization.

The Conference reached the following conclusions and recommendations:

- Many important economic and social decisions are being made today on long-term projects—major water resource management activities such as irrigation and hydropower; drought relief; agricultural land use; structural designs and coastal engineering projects; and energy planning—all based on the assumption that past climatic data, without modification, are a reliable guide to the future. This is no longer a good assumption since the increasing concentrations of greenhouse gases are expected to cause a significant warming of the global climate in the next century. It is a matter of urgency to refine estimates of future climate conditions to improve these decisions.
- Climate change and sea level rises due to greenhouse gases are closely linked with other major environmental issues, such as acid deposition and threats to the earth's ozone shield, mostly due to changes in the composition of the atmosphere by man's activities. Reduction of coal and oil use and energy conservation undertaken to reduce acid deposition will also reduce emissions of greenhouse gases; a reduction in the release of chlorofluorocarbons (CFCs) will help protect the ozone layer and will also slow the rate of climate change.
- While some warming of climate now appears inevitable due to past actions, the rate and degree of future warming could be profoundly affected by governmental policies on energy conservation, use of fossil fuels, and the emission of some greenhouse gases.

These conclusions are based on the following consensus of current basic scientific understanding:

- The amounts of some trace gases in the troposphere, notably carbon dioxide (CO_2), nitrous oxide (N_2O), methane (CH_4), ozone (O_3), and chlorofluorocarbons (CFC), are increasing. These gases are essentially transparent to incoming short-wave solar radiation but they absorb and emit long-wave radiation and are thus able to influence the earth's climate.
- The role of greenhouse gases other than CO_2 in changing the climate is already about as important as that of CO_2. If present trends continue, the combined concentrations of atmospheric CO_2 and other greenhouse gases would be radiatively equivalent to a doubling of CO_2 from preindustrial levels possibly as early as the 2030s.
- The most advanced experiments with general circulation models of the climatic system show increases of the global mean equilibrium surface

temperature for a doubling of the atmospheric CO_2 concentration, or equivalent, of between 1.5 and 4.5°C. Because of the complexity of the climatic system and the imperfections of the models, particularly with respect to ocean-atmosphere interactions and clouds, values outside this range cannot be excluded. The realization of such changes will be slowed by the inertia of the oceans; the delay in reaching the mean equilibrium temperatures corresponding to doubled greenhouse gas concentrations is expected to be a matter of decades.

- While other factors such as aerosol concentrations, changes in solar energy input, and changes in vegetation may also influence climate, the greenhouse gases are likely to be the most important cause of climate change over the next century.

- Regional scale changes in climate have not yet been modeled with confidence. However, regional differences from the global averages show that warming may be greater in high latitudes during late autumn and winter than in the tropics; annual mean runoff may increase in high latitudes; and summer dryness may become more frequent over the continents at middle latitude in the Northern Hemisphere. In tropical regions, temperature increases are expected to be smaller than the average global rise, but the effects on ecosystems and humans could have far-reaching consequences. Potential evapotranspiration probably will increase throughout the tropics whereas in moist tropical regions convective rainfall could increase.

- It is estimated on the basis of observed changes since the beginning of this century that global warming of 1.5°C to 4.5°C would lead to a sea-level rise of 20–140 centimeters. A sea-level rise in the upper portion of this range would have major direct effects on coastal areas and estuaries. A significant melting of the West Antarctic ice sheet leading to a much larger rise in sea level, although possible at some future date, is not expected during the next century.

- Based on analyses of observational data, the estimated increase in global mean temperature during the last one hundred years of between 0.3 and 0.7°C is consistent with the projected temperature increase attributable to the observed increase in CO_2 and other greenhouse gases, although it cannot be ascribed in a scientifically rigorous manner to these factors alone.

- Based on evidence of effects of past climatic changes, there is little doubt that a future change in climate of the order of magnitude obtained from climate models for a doubling of the atmospheric CO_2 concentration could have profound effects on global ecosystems, agriculture, water resources, and sea ice.

RECOMMENDED ACTIONS

- Governments and regional intergovernmental organizations should take into account the results of this assessment (Villach 1985) in their policies on social and economic development, environmental programs, and control of emissions of radiatively active gases.
- Public information efforts should be increased by international agencies and governments on the issues of greenhouse gases, climate change, and sea level, including wide distribution of the documents of this Conference (Villach 1985).
- Major uncertainties remain in predictions of changes in global and regional precipitation and temperature patterns. Ecosystem responses are also imperfectly known. Nevertheless, the understanding of the greenhouse question is sufficiently developed that scientists and policymakers should begin an active collaboration to explore the effectiveness of alternative policies and adjustments. Efforts should be made to design methods necessary for such collaboration.

  1. Governments and funding agencies should increase research support and focus efforts on crucial unsolved problems related to greenhouse gases and climate change. Priority should be given to national and international scientific program initiatives such as (a) the World Climate Research Program (WMO-ICSU), (b) present and proposed efforts on biogeochemical cycling and tropospheric chemistry in the framework of the Global Change Program proposed by ICSU, (c) National Climatic Research Program. Special emphasis should be placed on improved modeling of the ocean, cloud-radiation interactions, and land surface processes.

  2. Support for the analysis of policy and economic options should be increased by governments and funding agencies. In these assessments the widest possible range of social responses aimed at preventing or adapting to climate change should be identified, analyzed, and evaluated. These assessments should be initiated immediately and should employ a variety of available methods. Some of these analyses should be undertaken in a regional context to link available knowledge with economic decision-making and to characterize regional vulnerability and adaptability to climatic change. Candidate regions may include the Amazon Basin, the Indian subcontinent, Europe, the Arctic, the Zambezi Basin, and the North American Great Lakes.

- Governments and funding institutions should strongly support the following:

  1. Long-term monitoring and interpretation with state-of-the-art models of:

     - Radiatively important atmospheric constituents in addition to CO_2, including aerosols
     - Solar irradiance
     - Sea level

  2. Study and interpretation of the past history of climate and environment, specially regarding interactions among the atmosphere, oceans, and ecosystems.
  3. Studies of the effects of atmospheric composition and of changing climate and climatic extremes on subtropical and tropical ecosystems, boreal forests, and on water regimes.
  4. Investigations of the sensitivity of the global agricultural resource base with respect to:

     - Direct effects of increases in atmospheric CO_2 and other greenhouse gases
     - Effects of changes in climate
     - Probable combinations of these

  5. Evaluation of social and economic impacts of sea level rises.
  6. Analysis of policymaking procedures under the kinds of risks implied by a significant greenhouse warming.

- UNEP, WMO, and ICSU should establish a small task force on greenhouse gases, or take other measures, to:

  1. Help ensure that appropriate agencies and bodies follow up the recommendations of Villach 1985.
  2. Ensure periodic assessments are undertaken of the state of scientific understanding and its practical implications.
  3. Provide advice on further mechanisms and actions required at the national or international levels.
  4. Encourage research in developing countries to improve energy efficiency and conservation.
  5. Initiate, if deemed necessary, consideration of a global convention.

PART II

GLOBAL WARMING: BIOTIC SYSTEMS

Chapter 5

BIOTIC CAUSES AND EFFECTS OF THE DISRUPTION OF THE GLOBAL CARBON CYCLE

George M. Woodwell

The accumulation of heat-trapping gases in the atmosphere is the result of a disruption of the global carbon cycle by human activities. Two factors are clearly important: the massive conversion of fossil fuels into carbon dioxide, heat, and water, and the global destruction of forests, which results in the conversion of carbon stored in plants and soils to carbon dioxide, heat, and water. A third change has been suggested as the result of the warming itself: the mobilization by decay of carbon stored in soils and forests through the stimulation of respiration. This third source is potentially large—indeed, it could easily approach or exceed the magnitude of the current release from deforestation. Evidence is accumulating that it is active and not balanced by an equivalent stimulation of carbon storage due to the warming. It provides a positive feedback that makes stopping the further accumulation of heat-trapping gases more difficult as the warming progresses.

The global carbon cycle is sufficiently complex that there is reason to wonder how the atmosphere has remained as stable in composition as it appears to have been throughout the recent past. Scientists recognize that over the millennia the carbon dioxide content of the atmosphere is determined by the equilibrium established with the oceans, which have a vast capacity for storing carbon dioxide as carbon in various forms. The rate of exchange with the oceans is slow under most circumstances, however, and the emissions of carbon dioxide into the atmosphere from fossil fuels and from deforestation are rapid by comparison with the rate of transfer to the oceans. The result is the accumulation of carbon dioxide in the atmosphere that we are observing.

The problem is made more complex by rapid exchanges of carbon between the atmosphere and the terrestrial biota, especially forests. The importance of these exchanges is seen most clearly in the data of Charles David Keeling of the Scripps Institution of Oceanography. In 1958, with the encouragement of Roger Revelle, then director of Scripps, Keeling started the first detailed measurements of the concentration of carbon dioxide in the atmosphere with a now famous series of measurements made on Mauna Loa, a 13,000-foot peak in the Hawaiian Islands. Keeling's data, shown in Figure 5.1, reveal the upward trend in the concentration of carbon dioxide year by year, the various inflections in that trend (largely unexplained), and an annual oscillation with an amplitude of about 5 ppm. The oscillation shows a peak concentration during the late winter or spring, usually in April, and a minimum in early October. It is considerably greater at higher latitudes in the Northern Hemisphere, less at lower latitudes, and is reversed in its seasonality in the Southern Hemisphere. The oscillation is caused by the metabolism of terrestrial ecosystems, especially forests. The amplitude is greater in the Northern Hemisphere because the land area is larger and forests are more abundant. The oscillation is important in this discussion only in emphasizing that terrestrial ecosystems and forests in particular affect the composition of the atmosphere in a period as short as weeks. The emphasis is on forests because they are large in area, large in stature, carry on most of the photosynthesis and respiration on land, and store carbon compounds seasonally and from year to year. Moreover, their metabolism controls an amount of carbon stored in

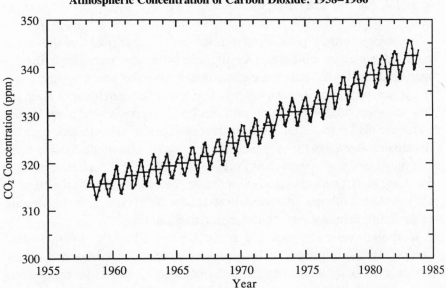

FIGURE 5.1
Atmospheric Concentration of Carbon Dioxide: 1958–1986

Data from the Mauna Loa observatory according to the record maintained by C. D. Keeling and his associates since 1958. The upward trend has increased over the past 10 years to about 1.5 ppmv annually. The causes of the inflections in the upward trend have not been defined. The oscillation reaches a peak in April and a minimum in late September or early October. The oscillation is due to the metabolism of the vegetation of the Northern Hemisphere—especially forests which store carbon during the summer and release it through respiration in winter. The oscillation is conspicuous because the cycles of storage and release are out of phase.

trees and soils that exceeds the approximately 700 billion tons in the atmosphere. Thus a small change in their metabolism can affect the composition of the atmosphere significantly.

Our purpose is to define the interactions between the biota and the atmosphere as the earth warms in response to the accumulation of heat-trapping gases. The biotic interactions are large and important enough to warrant early action in stabilizing the composition of the atmosphere, a step that becomes more difficult as effects accumulate.

EFFECTS OF WARMING: BIOTIC SOURCES OF CARBON

Climatologists are virtually unanimous in their appraisal that the dominant influence on climates globally over the next decades will be a warming due to the increasing concentrations of the heat-trapping gases in the atmosphere. The warming is due to carbon dioxide and methane in particular, although ozone and the CFCs, nitrous oxide, and other gases are also implicated. The influence of carbon dioxide and methane dominates because these gases are the most abundant of the heat-trapping trace gases, and their concentrations are accumulating rapidly. Methane is disproportionately important because its absorption band in the infrared is otherwise unobstructed and a small change in the amount of methane brings a large change in temperature.

Carbon dioxide and methane are both involved in the global cycle of carbon. Both are products of respiration, the process by which living systems break down carbon compounds, turning them into energy and water as well as carbon dioxide, methane, or a limited number of other organic products. For our purposes we can think of respiration as either aerobic or anaerobic. Aerobic respiration dominates when oxygen is freely available; anaerobic, when oxygen is limited. Aerobic respiration produces carbon dioxide; anaerobic respiration, methane.

A further set of data is basic to consideration of the role of the biota. A comparison of the amount of carbon held in the atmosphere with the amount held in various biotically controlled pools shows that there is approximately three times more held in the biotic pools (Figure 5.2) than is held in the atmosphere. This relationship means that small changes in rates of exchange with the atmosphere or among the biotically controlled pools have the potential for affecting the composition of the atmosphere significantly. If, for instance, there is a change in the area of forests due to the expansion or contraction of agriculture, we can expect the flow of carbon to or from forests to change. The magnitude of that change can be large enough to affect the composition of the atmosphere significantly. That is, of course, what is happening: rapid deforestation in the tropics is resulting in a surge in the release of carbon from forests into the atmosphere. The magnitude of the release is uncertain. Current estimates are that the net release from deforestation

FIGURE 5.2
Exchangeable Carbon Reservoirs and Fluxes

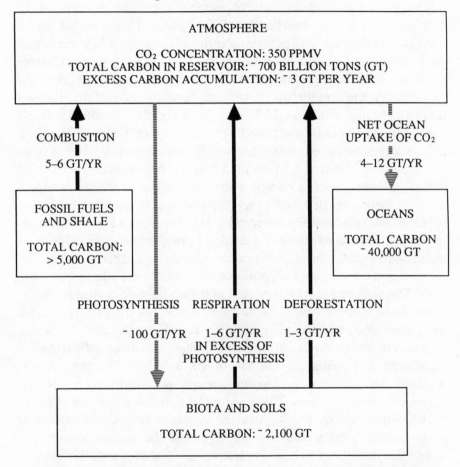

The fluxes are in billions of metric tons of carbon per year. The size of the carbon reservoirs is in billions of metric tons of carbon. Carbon transferred to the atmosphere by respiration exceeds the carbon being fixed by photosynthesis because global warming increases the rate of respiration more than it increases the rate of photosynthesis, although the size of the increase is uncertain.

globally is 1 to 3 billion tons annually. That release is to be compared with the nearly 6 billion tons of carbon released through combustion of fossil fuels.

The amount of warming to be expected is uncertain. A recent set of data prepared by James Hansen of the Goddard Institute of Space Studies at Columbia, and presented at the 31 June 1988 hearing before the U.S. Senate Committee on Energy and Natural Resources (Chapter 2), shows that the earth has warmed between 0.5 and 0.7°C over the past century and that the warming is accelerating. Five of the warmest years (including 1988) in 130 years have occurred in the 1980s. If present trends continue and no effort is made to limit deforestation and reduce dependence on fossil fuels, the warming may approach an average for the earth as a whole of 0.5 to 1°C per decade over the next several decades. Such a rate of warming implies substantially higher rates, perhaps two or more times the average, in middle and higher latitudes and substantially lower rates, perhaps half or less the average, in low latitudes. Such rates of warming in mid-latitudes are especially important because these zones contain large areas that are normally forested and contain large amounts of carbon in trees, other plants, and soils. The warming can be expected to bring substantial changes in the vigor of these forests throughout their ranges, but the effects become most clear along their warmer and drier margins.

In the middle latitudes currently a 1°C change in mean temperature is equivalent to a change in latitude of 100 to 150 kilometers. A 1°C warming could be expected to cause the migration of the forest–prairie border by such an amount. There would be a simultaneous migration of the transitions among forest types: the transition from deciduous forest to the boreal forest would migrate, as would the various forest types within the deciduous forest. Such a series of changes in the circumstances of forests would be highly destructive, the more so as the speed of the changes increased. Trees are easily killed; forests regenerate only if conditions remain favorable for decades or longer. The effect of rapid warming, measured here as in excess of 0.1°C per decade, is the destruction of forests over large areas with the release of substantial quantities of carbon into the atmosphere, a positive feedback that makes the problem worse.

THE GLOBAL CARBON BALANCE:
BIOTIC INFLUENCES

The annual oscillation in the concentration of carbon dioxide observed in the Mauna Loa data of Figure 5.1 is the result of an annual shift in the metabolism of forests. Net photosynthesis dominates in summer and removes carbon dioxide from the atmosphere; respiration dominates in winter and releases carbon dioxide into the atmosphere. The oscillation is merely the net of two opposing processes that actually involve very much larger quantities of carbon.

The total amount of carbon in the atmosphere is in excess of 700 billion tons. The annual absorption of carbon by plants globally through photosynthesis is about 100 billion tons, approximately one-seventh of the atmospheric burden. There is an approximately equal release of carbon from the biota and soils into the atmosphere through respiration. Any factor that affects this exchange, either the photosynthetic absorption or the release through respiration, has the potential for affecting the amount of carbon in the atmosphere significantly. What factors might affect it?

The answer reaches to the core of one of the most threatening aspects of climatic change. Yet the answer is not clear and probably will not be clear until long after the effects of a climatic change are well under way. Photosynthesis is affected by many factors. Chief among them is the availability of energy as light, the presence of chlorophyll, and the availability of water and nutrients. Respiration too is affected by many factors including the availability of water and an appropriate temperature. Respiration is particularly sensitive to temperature: a 1°C increase in temperature often increases the rate of respiration by 10 to 30% and sometimes more. No such sensitivity to temperature exists for photosynthesis. That observation alone suggests that a global warming will speed the decay of organic matter globally without affecting photosynthesis appreciably. There is enough carbon available in soils and forests—especially in the middle and higher latitudes where the warming is expected to be greatest—to affect the composition of the atmosphere significantly. If, for example, we assume that a 0.5°C warming has occurred globally and that in the forested middle latitudes the warming approaches two times the mean for the earth as a whole and is

0.75 to 1.0°C, then we might be experiencing in these latitudes an increase in respiration of between 5 and 20%. The increase might apply to 20 to 30% of the global total of respiration on land or 20 to 30 billion tons of carbon. The excess release from the warming already experienced would be 1 to 6 billion tons of carbon per year, if there were no compensation through increased photosynthesis.

Is there any basis for expecting an increase in the photosynthetic storage of carbon—a compensatory stimulation of the fixation of carbon by plants and storage in forests and soils? Will not the warming, for example, open new lands in high latitudes to forests and result in the spread of forests into regions now tundra? Perhaps that will occur, but the difficulty lies in the speed of the transitions. Forests require decades or centuries to develop, especially on lands where soils are thin and nutrients are in short supply. They also require a climatic stability, sources of seeds, and the development of soils and the accumulation of sufficient stocks of nutrients in soils and plants to support a forest. The climatic transitions under way at the moment, unless deliberately checked, are rapid by any measure and are expected to be continuous into the indefinite future. They do not offer conditions under which forests develop on new lands and store carbon for long periods.

Is it possible that existing forests or tundra will be stimulated to store additional carbon in plants and soils as the warming progresses? Perhaps. The boreal forest and other coniferous forests may be sufficiently resilient to respond to a warming by increased periods of photosynthesis. Whether the carbon fixed will be stored or simply released through increased rates of respiration remains an open question. There is also the possibility that the tundra, warmed, will respond in surprising ways, including an increase in primary production and increased storage of carbon in peat. Much will hinge on the availability of water. A wetter tundra might store additional carbon in soils; a drier tundra might release it through the decay of organic soils long frozen or normally frozen throughout much of the year.

The overall response of the earth to rapid warming seems most likely to be through a net stimulation of total respiration above gross photosynthesis with a net additional release of carbon into the atmosphere as the earth warms. The magnitude of the release will hinge heavily on the rate of warming: the more rapid the warming, the more severe the release.

Yet the problem is even more complex and threatening: there is an increase in the amount of methane in the atmosphere. This increase is rapid—about 1% per year—and significant because methane is, molecule for molecule, more effective than carbon dioxide in trapping heat. Methane is a product of respiration in places where oxygen is limited. The major source of methane globally is probably soils, especially wet soils, including marshes, swamps, and bogs. The global warming that has already occurred has without question stimulated anaerobic decay as well as aerobic decay and is the dominant source of the increased rate of release of methane. While the observation is far from proved, it is sufficiently obvious to ecologists that it provides another reason for concern that we are observing biotic feedbacks from the warming that has already occurred.

IS THERE A THRESHOLD FOR BIOTIC EFFECTS?

This discussion implies that there must be a threshold for effects of a warming on the global biota, especially with respect to efforts to stabilize fluxes of carbon through the biota. The fact is there are few thresholds in nature. Changes in the structure of nature tend to be continuous as environmental factors change. Thresholds are introduced as a human convenience, often inappropriately. What is the threshold for safe use of a substance such as DDT or the PCBs that can be accumulated in fish, birds, and mammals (including people) to concentrations that are a millionfold higher than the concentrations in soil or water? Just how fast can we allow the earth to warm and move climates out from under the present order of things, out from under our present agriculture, our present pattern of water availability and use, our present fisheries, out from under our present civilization?

Obviously, no change in climate is desirable on an earth that is increasingly crowded with an expanding human enterprise. If a warming is inevitable but its rate can be controlled, we might seek guidance in estimating a rate that is acceptable. What rate might be accommodated by forests and other terrestrial ecosystems? We have experienced a warming of between 0.5 and 0.7°C over the past century. The consequences of that warming have not yet been fully recognized. One consequence not yet resolved is the increased rate of accumulation of methane in the atmosphere. Nonetheless, we might introduce a margin of safety of a factor of 2—a modest factor in dealing with the safety of the whole earth—and decide more or less arbitrarily that a global average rate of warming in excess of 0.25°C per century should be avoided. But an average change in the temperature of the earth is not the maximum change. Changes in the middle and high latitudes exceed the average by a factor in excess of 2. Thus an objective of slowing the warming to a rate less than 0.25°C per century seems essential to preserving forests and avoiding the danger that the warming will speed itself through the destruction of forests and soils in middle and high latitudes.

WHAT CAN BE DONE?

The objective becomes stabilization of the temperature of the earth, at least insofar as the temperature is being affected by the heat-trapping gases. Stabilization will require as the first step the stabilization of the composition of the atmosphere. The current rate of accumulation of carbon in the atmosphere is about 1.5 ppm per year or about 3.0 billion tons. The actual amount transferred into the oceans is not known; nor do we know the actual total release of carbon into the atmosphere. The total release is made up of several contributions: the release from burning fossil fuels (thought to be about 5.6 billion tons of carbon annually); a further release from deforestation (1 to 3 billion tons); and a further release of unknown magnitude from the stimulation of respiration due to the warming itself. This latter source also results in an increase in the amount of methane as anaerobic respiration is stimulated globally.

The important point at the moment, however, is that the annual accumulation of carbon in the atmosphere is known to be about 3 billion tons of carbon as carbon dioxide. Because there is an additional accumulation from other gases such as methane whose emissions cannot be controlled yet, a further restriction in carbon dioxide emissions is necessary to compensate. We might seek to remove at least 3 to 4 billion tons of carbon from current emissions. Such a step is possible.

Three separate actions are needed. There is no alternative to a shift away from reliance on fossil fuels as the primary source of energy. A 50% reduction in use of fossil fuels globally below the current use would reduce current emissions by about 2.8 billion tons. A cessation of deforestation globally would reduce emissions by a further 1 to 3 billion tons. Reforestation of about 2 million square kilometers would result in the storage of about 1 billion tons of carbon annually in the trees and soils. That rate of storage would continue for 40 to 60 years until the forest reached stability and total respiration equaled gross photosynthesis. Not one of these steps alone is adequate. But all are desirable, commonsense steps toward wise management of the human habitat quite apart from the need to stabilize climates.

A shift away from fossil fuels will require greatly improved efficiency of energy use to provide needed goods and services; it will also require a strong commitment to renewable energy sources, primarily solar. The use of biomass is a promising way to capture and use solar energy. If this biomass were to come from existing forests, it would result in even larger transfers of carbon from biotic pools into the atmosphere and hence increase the rate and extent of global warming. If, however, land not now in forests were used for energy plantations for fast-growing trees or other crops, an equal amount of carbon would be removed from the atmosphere as the energy crops were grown as would be returned when they were burned as fuel. The use of biomass in this form would then contribute to the solution rather than exacerbate the global warming.

The challenge of climatic change is complex and difficult, but not impossible. A solution will require global progress in management of resources. It will require a global convention dealing with the atmosphere and energy and accommodating aspirations for technological and economic development. It will also have to provide for reasoned management of the world's forests—a step long overdue for reasons quite apart from the global climatic changes now under way. There are many uncertainties in our knowledge of the causes of the changes in climate, their magnitude, and the effects. These uncertainties simply emphasize that in treating cause and effect in experiments involving the whole earth, we are working at or beyond the limits of human knowledge and surprises are common—sometimes unpleasant ones as large and serious as the ozone hole. Reason argues for a studied conservatism in such experiments.

Chapter 6

EFFECTS OF GLOBAL WARMING ON BIOLOGICAL DIVERSITY

Robert L. Peters

ABSTRACT

Previous natural climate changes have caused large-scale geographical shifts, changes in species composition, and extinctions among biological communities. If the widely predicted greenhouse effect occurs, communities would respond in similar ways. Moreover, population reduction and habitat destruction due to other human activities would make it difficult for species to shift ranges in response to changing climate conditions. Human encroachment and climate change would jointly threaten many more species than either alone.

From the Proceedings of the First North American Conference on Preparing for Climate Change: A Cooperative Approach, Washington, D.C., 27–29 October 1987.

INTRODUCTION

If the planet warms as projected, natural ecosystems would be stressed by large changes in temperature, moisture patterns, evaporation rates, and other associated chemical and physical changes.

We can infer how the biota might respond by observing the world as it is today. By observing present distributions of plants and animals, which are determined in large part by temperature and moisture patterns, it is possible to hypothesize what would happen if these underlying temperature and moisture patterns changed.

For example, if we know that one race of the dwarf birch, *Betula nana,* can only grow where the temperature never exceeds 22°C (Ford, 1982), then we can hypothesize that it would disappear from those areas where global warming causes temperatures to exceed 22°C.

Ecologists can also observe the results of the many small climate experiments performed by nature every year. We can observe what happens to the birch trees if unusually warm weather occurs during a single year. Perhaps some trees fail to set seed. What happens if there are several warm years in a row? Perhaps some individual trees die.

Also, scientists can look into the past to see how the ranges of plants and animals varied in response to past climate change. A palynologist can count the types of plant pollen found at different depths in the soil, each depth corresponding to the time in which a particular layer of soil was laid down. If birch pollen is found at a depth corresponding to 10,000 years ago, birch trees must have lived in the area at that time.

The fossils of animals convey similar information. If we find tapir and peccary bones in Massachusetts' sediments corresponding to a previous interglacial period of warming, as has been done (Dorf, 1976), we can infer from the presence of the bones that regional temperatures were then higher in Massachusetts. Better indications can be gotten from the fossil bones of small mammals, like rodents, which generally spend their lives within a small area and therefore match the local habitat very well.

These kinds of observations tell us that plants and animals are very sensitive to climate. Their ranges move when the climate patterns change—species die out in areas where they were once found and colonize new areas where the climate becomes newly suitable. We also know from the fossil record that some species have become completely extinct because they were unable to find suitable habitat when climate change made their old homes unlivable.

As I will discuss below, there will be many ways in which climate change will stress and change natural ecosystems. If warming occurs as projected, during the next 50 years it is likely to change the ranges of many species, disrupt natural communities, and contribute to the extinction of species.

THE NATURE OF THE ECOLOGICALLY SIGNIFICANT CHANGES

There is widespread consensus that global warming will occur during the next century and that a global warming of 3°C may be reached during the next 50 years (Hansen et al., 1981; NRC, 1983; Schneider and Londer, 1984). Ecologically significant temperature rise would occur during the transitional warming phase, well before 3°C is reached—as discussed below, warming of less than 1°C may have substantial ecological effects.

For the purpose of discussion in this paper, I will take average global warming to be 3°C, but it must be recognized that additional warming well beyond 3°C may be reached during the next century if the production of anthropogenic greenhouse gases continues.

The threats to natural systems are serious for the following reasons. First, three degrees of warming would present natural systems with a warmer world than has been experienced in the past 100,000 years (Schneider and Londer, 1984). This warming would not only be large compared to recent natural fluctuations, but it would be very fast, perhaps an order of magnitude faster than past natural changes. For reasons discussed in detail below, such a rate of change may prove more than many species can adapt to. Moreover, human encroachment—habitat destruction—will make wild populations small and vulnerable to local climate changes.

Second, ecological stress would not be caused by temperature rise alone. Changes in global temperature patterns would trigger widespread alterations in rainfall patterns (Hansen et al., 1981; Kellogg and Schware, 1981; Manabe et al., 1981; Wigley et al., 1980), and we know that for many species precipitation is a more important determinant of survival than temperature per se. Some regions would see dramatic increases in rainfall, and others would lose their present vegetation because of drought. For example, Kellogg and Schware (1981), drawing on knowledge of past vegetation patterns, project substantial decreases in rainfall in America's Great Plains—perhaps as much as 40% by the early decades of the next century.

Other environmental factors would change because of global warming: Soil chemistry would change (Kellison and Weir, 1986). Increased carbon dioxide concentrations may accelerate the growth of some plants at the expense of others (NRC, 1983; Strain and Bazzaz, 1983), possibly destabilizing natural ecosystems. And rises in sea level may inundate coastal biological communities (NRC, 1983; Hansen et al., 1981; Hoffman, Keyes, and Titus, 1983; Titus et al., 1985). What all this means is that the ranges of individual species would shift and that ecological systems would be disrupted.

One important pattern of global warming, generally concluded by a variety of computer projections, is that warming will be relatively greater at higher latitudes. This suggests that, although tropical systems may be more diverse and are currently under great threat because of habitat destruction, temperate zone and arctic species may ultimately be in greater jeopardy from climate change. Also, a recent attempt to map climate-induced changes in world biotic communities projects that high-altitude communities would be particularly stressed (Emanuel et al., 1985). Boreal forest, for example, was projected to decrease by 37% in response to warming of 3°C.

HOW DO SPECIES RESPOND TO WARMING?

We know that when temperature and rainfall patterns change, species ranges change. Even very small temperature changes of less than one degree within this century have been observed to cause substantial range changes. For example, the white admiral butterfly (*Ladoga camilla*) and the comma butterfly (*Polygonia calbum*) greatly expanded their ranges in the British Isles during the past century as the climate warmed approximately 0.5°C (see in Ford, 1982). At the same time, other species that depend upon cooler conditions, like the ant *Formica lugubris,* retracted their ranges into the cooler uplands.

On a larger ecological and temporal scale, entire vegetation types have shifted in response to past temperature changes no larger than those that may occur during the next 100 years or less (Baker, 1983; Bernabo and Webb, 1977; Butzer, 1980; Flohn, 1979; Muller, 1979; Van Devender and Spaulding, 1979). Such shifts show general patterns. As the earth warms, species tend to shift to higher latitudes and altitudes. From a simplified point of view, rising temperatures have caused species to colonize new habitats toward the poles, often while their ranges contracted away from the equator as conditions there became unsuitable.

During several Pleistocene interglacials, for example, the temperature in North America was apparently 2° to 3°C higher than now. Osage oranges and pawpaws grew near Toronto, several hundred kilometers north of their present distribution; manatees swam in New Jersey; tapirs and peccaries foraged in North Carolina (Dorf, 1976). Other significant changes in species ranges have been caused by altered precipitation accompanying past global warming, including expansion of prairie in the American Midwest during a global warming episode approximately 7,000 years ago (Bernabo and Webb, 1977).

It should not be imagined, because species tend to shift in the same general direction, that existing biological communities move in synchrony. Conversely, because species shift at different rates in response to climate change, communities often disassociate into their component species. Recent studies of fossil packrat (*Neotoma spp.*) middens in the southwestern United States show that during the wetter, moderate climate of 22,000–12,000 years ago, there was not a concerted shift of communities. Instead, species responded individually to climatic change, forming stable, but by present-day standards unusual, assemblages of plants and animals (Van Devender and Spaulding, 1979). In eastern North America, too, postglacial communities were often ephemeral associations of species, changing as individual ranges changed (Davis, 1983).

A final aspect of species response is that species may shift altitudinally as well as latitudinally. When climate warms, species shift upward. Generally, a short climb in altitude corresponds to a major shift in latitude: the 3°C cooling of 500 meters in elevation equals roughly

250 kilometers in latitude (MacArthur, 1972). Thus, during the middle Holocene, when temperatures in eastern North America were 2°C warmer than at present, hemlock (*Tsuga canadensis*) and white pine (*Pinus strobus*) were found 350 meters higher on mountains than they are today (Davis, 1983).

Because mountain peaks are smaller than bases, as species shift upward in response to warming, they typically occupy smaller and smaller areas, have smaller populations, and may thus become more vulnerable to genetic and environmental pressures. Species originally situated near mountaintops might have no habitat to move up to and may be entirely replaced by the relatively thermophilous species moving up from below. Examples of past extinctions attributed to upward shifting include alpine plants once living on mountains in Central and South America, where vegetation zones have shifted upward by 1,000–1,500 m since the last glacial maximum (Flenley, 1979; Heusser, 1974).

MAGNITUDE OF PROJECTED LATITUDINAL SHIFTS

Although Pleistocene and past Holocene warming periods were probably not due to elevated CO_2 levels, researchers have predicted that, if the proposed CO_2-induced warming occurs, similar species shifts would also occur, and vegetation belts would move hundreds of kilometers toward the poles (Frye, 1983; Peters and Darling, 1985). Three hundred kilometers of shifting in the temperate zone is a reasonable estimate for a 3°C warming, based on the positions of vegetation zones during analogous warming periods in the past (Dorf, 1976; Furley et al., 1983).

Additional confirmation that shifts of this magnitude may occur comes from attempts to project future range shifts for some species by looking at their ecological requirements. For example, the forest industry is concerned about the future of commercially valuable North American species, like the loblolly pine (*Pinus taeda* L.). This species is limited to the south of its range by moisture stress on seedlings. Based on these physiological requirements for temperature and moisture, Miller, Dougherty, and Switzer (1987) projected a range retraction to the north of approximately 350 kilometers in response to a global warming of 3°C.

DISPERSAL RATES AND BARRIERS

The ability of species to adapt to changing conditions will to a large extend depend upon their ability to track shifting climatic optima by dispersing colonists. In the case of warming, a North American species, for example, would most likely need to establish colonists to the north. Survival of plants and animals would therefore depend either upon long-distance dispersal of colonists, such as seeds or migrating animals, or upon rapid iterative colonization of nearby habitat until long-distance shifting is accomplished. If a species' intrinsic dispersal rate is low, or if barriers to dispersal are present, extinction may result.

There are many cases where complete or local extinction has occurred because species were unable to disperse rapidly enough when climate changed. For example, a large, diverse group of plant genera, including water-shield (*Brassenia*), sweet gum (*Liquidambar*), tulip tree (*Liriodendron*), magnolia (*Magnolia*), moonseed (*Menispermum*), hemlock (*Tsuga*), arbor vitae (*Thuja*), and white cedar (*Chamaecyparis*), had a circumpolar distribution in the Tertiary. But during the Pleistocene ice ages, all went extinct in Europe while surviving in North America. Presumably, the east-west orientation of such barriers as the Pyrenes, Alps, and the Mediterranean, which blocked southward migration, was partly responsible for their extinction (Tralau, 1973).

Other species thrived in Europe during the cold periods, but could not survive conditions in postglacial forests. Some were unable to extend their ranges northward in time and became extinct except in cold, mountaintop refugia (Seddon, 1971).

These natural changes were comparably slow: Change to warmer conditions at the end of the last ice age spanned several thousand years, yet is considered rapid by natural standards (David, 1983). We can deduce that, if such a slow change was too fast for many species to adapt to, the projected warming—perhaps ten times faster—will have more severe consequences. For widespread, abundant species, like the loblolly pine modeled by Miller, Dougherty, and Switzer (1987), even substantial range retraction might pose little threat of extinction; but rare, localized species, whose entire ranges might become unsuitable, would be threatened unless dispersal and colonization were successful.

Could an average species successfully disperse given what we know about dispersal rates and barriers? If the climatic optima of temperate zone species do shift hundreds of miles toward the poles within the next 100 years, then these species would have to colonize very rapidly. A localized species might have to shift poleward at several hundred kilometers per century, or faster, in order to avoid being left behind in areas too warm for survival. Although some species, such as plants propagated by spores or dust seeds, may be able to match these rates (Perring, 1965), many species could not disperse fast enough to compensate for the expected climatic change without human assistance (see in Rapoport, 1982), particularly given the presence of dispersal barriers. Even wind-assisted dispersal may fall short of the mark for many species. For example, wind scatters seeds of the grass *Agrostis hiemalis,* but 95% fall within 9 m of the parent plant (Willson, 1983). In the case of the Engelmann spruce, a tree with light, wind-dispersed seeds, fewer than 5% of seeds travel even 200 m downwind, leading to an estimated migration rate of between 1 and 20 km per century (Seddon, 1971).

Figure 6.1 illustrates the difficulties to be faced by a population whose habitat becomes unsuitable due to climate change. Colonists (e.g., seeds) must run an obstacle course through various natural and human-created dispersal barriers in a limited amount of time in order to reach habitat that will be suitable under the new climatic regime. For the example selected, with a dispersal rate of 20 km per century, successful dispersal is highly improbable.

Although many *animals* may be, in theory, highly mobile, the distribution of some is limited by the distributions of particular plants; their dispersal rates may therefore largely be determined by those of co-occurring plants. Behavior may also restrict dispersal even of animals physically capable of large movements. Dispersal rates below 2.0 km/year have been measured for several species of deer (Rapoport, 1982), and many tropical deep-forest birds simply do not cross even very small unforested areas (Diamond, 1975). On the other hand, some highly mobile animals, particularly those whose choice of habitat is relatively unrestrictive, may shift rapidly. Several authors (see Edgell, 1984) have suggested, for instance, that climate change caused major range shifts in some European migratory waterfowl in this century.

Even if animals are good dispersers, suitable habitat may be reduced under changing climatic conditions. For example, it has been suggested that tundra nesting habitat for migratory shore birds might be reduced by high-arctic warming (Myers, 1988).

FIGURE 6.1
Obstacles Faced by Species in Responding to Climate Change

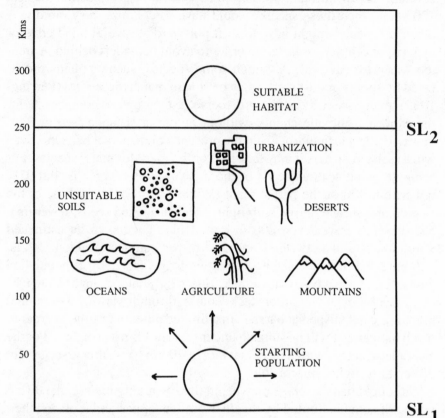

Obstacle course to be run by species facing climatic change in a human-altered environment. To "win," a population must track its shifting climatic optimum and reach suitable habitat north of the new southern limit of the species range. SL_1 = species southern range limit under initial conditions. SL_2 = southern limit after climate change. The model assumes a plant species consisting of a single population, which has its distribution determined solely by temperature. After a 3°C rise in temperature the population must have shifted 250 km to the north to survive, based on Hopkins' bioclimatic law (MacArthur, 1972). Shifting will occur by simultaneous range contraction from the south and expansion by dispersion and colonization to the north. Progressive shifting depends upon propagules that can find suitable habitat to mature and in turn produce propagules that can colonize more habitat to the north. Propagules must pass around natural and artificial obstacles like mountains, lakes, cities, and farm fields. The Englemann spruce has an estimated, unimpeded dispersal rate of 20 km/100 years (Seddon, 1971). Therefore, for this species to "win," colonizing habitat to the north of the shifted hypothetical limit would require a minimum of 1,250 years.

SYNERGY OF HABITAT DESTRUCTION
AND CLIMATE CHANGE

We know that even slow, natural climate change caused species to become extinct. What is likely to happen given the environmental conditions of the coming century?

Some clear implications for conservation follow from the preceding discussion of dispersal rates. Any factor that would decrease the probability that a species could successfully colonize new habitat would increase the probability of extinction.

Thus, species are more likely to become extinct if their remaining populations are small. Smaller populations mean fewer colonists can be sent out and that the probability of successful colonization is smaller.

Species are more likely to become extinct if they occupy a small geographic range. It is less likely that some part of their range will remain suitable when climate changes. Further, if a species has lost much of its range because of some other factor, like habitat destruction, it is possible that the remaining populations are not located in prime habitat—they might now be found in that part of their historic range which is most susceptible to climate change.

As previously described, species are more likely to become extinct if there are physical barriers to the colonization, such as oceans, mountains, and cities (Figure 6.1).

For many species, all of these conditions will be met by human-caused habitat destruction, which increasingly confines the natural biota to small patches of original habitat, patches isolated by vast areas of human-dominated urban or agricultural lands. This problem by itself threatens hundreds of thousands of plant and animal species with extinction within the next 20 years (Myers, 1979; Lovejoy, 1980).

Habitat destruction in conjunction with climate change sets the stage for an even larger wave of extinction than previously imagined, based upon consideration of human encroachment alone. Small, remnant populations of most species, surrounded by cities, roads, reservoirs, and farm land, would have little chance of reaching new habitat if climate change makes the old unsuitable. Few animals or plants would be able to cross Los Angeles on the way to the promised land.

WHAT THIS MEANS FOR MANAGEMENT

There is not space to detail management options in this paper. See Peters and Darling (1985) for a further discussion. In brief, however, there are possible strategies to mitigate species loss. Better characterization of future regional climatic regimes is vital. Such information could be used to make better decisions about siting or modifying reserves. Corridors, for example, particularly in mountainous regions where dispersal distances need not be large, could provide avenues for dispersal. Similarly, managers of coastal marshes might wish to ensure that uplands are conserved in anticipation of when rising sea level forces marshes upward. Forewarning of local environmental trends might also allow reserve managers to prepare for active management of reserve conditions, as by irrigating to compensate for decreased rainfall.

The most comprehensive conclusion, however, is that because species with fragmented populations and reduced ranges are so vulnerable to climate change, one of the best things that can be done now is to minimize further range reduction. Thus, climate change is a compelling new reason to conserve as many natural lands as possible.

CONCLUSION

Finally, not all the changes for wild systems would be negative. Some species would expand their ranges and have greater abundances. It has been suggested, for example, that some arctic-nesting waterfowl might expand their populations as conditions warm (Harington, 1986).

However, the most optimistic thing that should be said about the future of natural systems under a regime of warming climate is that a great deal of rearrangement would occur, and it is most likely the outcome will be widespread extinction of species.

ACKNOWLEDGMENTS

Please see Peters and Darling (1985) for a complete list of acknowledgments for help with this work. Many of the ideas presented here derive from that paper and from the contributions of Joan Darling.

References

Baker, R. G. 1983. Holocene vegetational history of the western United States. Pages 109–125 in H. E. Wright, Jr., ed., *Late-Quaternary Environments of the United States*. Volume 2. *The Holocene*. University of Minnesota Press: Minneapolis.

Bernabo, J. C., and T. Webb III. 1977. Changing patterns in the Holocene pollen record of northeastern North America: a mapped summary. *Quat. Res.* 8: 64–96.

Butzer, K. W. 1980. Adaptation to global environmental change. *Prof. Geogr.* 32(3): 269–278.

Davis, M. B. 1983. Holocene vegetational history of the eastern United States. Pages 166–181 in H. E. Wright, Jr., ed., *Late-Quaternary Environments of the United States*. Volume 2. *The Holocene*. University of Minnesota Press: Minneapolis.

Diamond, J. M. 1975. The island dilemma: lessons of modern biogeographic studies for the design of natural preserves. *Biol. Conserv.* 7: 129–146.

Dorf, E. 1976. Climatic changes of the past and present. Pages 384–412 in C. A. Ross, ed., *Paleobiogeography: Benchmark Papers in Geology 31*. Dowden, Hutchinson, and Ross: Stroudsburg, PA.

Edgell, M.C.R. 1984. Trans-hemispheric movements of Holarctic Anatidae: the Eurasian wigeon (*Anas penelope L.*) in North America. *J. Biogeogr.* 11: 27–39.

Emmanuel, W. R., H. H. Shugart, and M. P. Stevenson. 1985. Response to comment: "Climatic change and the broad-scale distribution of terrestrial ecosystem complexes." *Clim. Change* 7: 457–460.

Flenley, J. R. 1979. *The Equatorial Rain Forest:* Butterworths: London.

Flohn, H. 1979. Can climate history repeat itself? Possible climatic warming and the case of paleoclimatic warm phases. Pages 15–28 in W. Bach, J. Pankrath, and W. W. Kellogg, eds., *Man's Impact on Climate*. Elsevier Scientific Publishing: Amsterdam.

Ford, M. J. 1982. *The Changing Climate*. George Allen and Unwin: London.

Frye, R. 1983. Climatic change and fisheries management. *Nat. Resources J.* 23: 77–96.

Furley, P. A., W. W. Newey, R. P. Kirby, and J. McG. Hotson. 1983. *Geography of the Biosphere*. Butterworths: London.

Hansen, J., D. Johnson, A. Lacis, S. Lebedeff, P. Lee, D. Rind, and G. Russell. 1981. Climate impact of increasing atmospheric carbon dioxide. *Science* 213: 957–966.

Hansen, J., A. Lacis, D. Rind, G. Russell, I. Fung, and S. Lebedeff. 1987. Evidence for future warming: how large and when. In W. E. Shands and

J. S. Hoffman, eds., *CO₂, Climate Change, and Forest Management in the United States.* Conservation Foundation: Washington, D.C.

Harington, C. R. 1986. The impact of changing climate on some vertebrates in the Canadian arctic. In *Proceedings, Impact of Climate Change on the Canadian Arctic; Orillia, Ontario, March 3–5, 1986.* Atmospheric Environment Service, Downsview, Ont.

Heusser, C. J. 1974. Vegetation and climate of the southern Chilean lake district during and since the last interglaciation. *Quat. Res.* 4: 290–315.

Hoffman, J. S., D. Keyes, and J. G. Titus. 1983. *Projecting Future Sea Level Rise.* U.S. Environmental Protection Agency: Washington, D.C.

Kellison, R. C., and R. J. Weir. 1986. Selection and breeding strategies in tree improvement programs for elevated atmospheric carbon dioxide levels. In W. E. Shands and J. S. Hoffman, eds., *CO₂, Climate Change, and Forest Management in the United States.* Conservation Foundation: Washington, D.C., in press.

Kellogg, W. W., and R. Schware. 1981. *Climate Change and Society: Consequences of Increasing Atmospheric Carbon Dioxide.* Westview Press: Boulder, CO.

Lovejoy, T. E. 1980. A projection of species extinctions. Pages 328–331 in *The Global 2000 Report to the President: Entering the Twenty-first Century.* Council on Environmental Quality and the Department of State. U.S. Government Printing Office: Washington, D.C.

MacArthur, R. H. 1972. *Geographical Ecology.* Harper & Row: New York.

Manabe, S., R. T. Wetherald, and R. J. Stouffer. 1981. Summer dryness due to an increase of atmospheric CO_2 concentration. *Clim. Change* 3: 347–386.

Miller, W. F., P. M. Dougherty, and G. L. Switzer. 1986. Rising CO_2 and changing climate: major southern forest management implications. In W. E. Shands and J. S. Hoffman, eds., *CO₂, Climate Change, and Forest Management in the United States.* Conservation Foundation: Washington, D.C., in press.

Muller, H. 1979. Climatic changes during the last three interglacials. Pages 29–41 in W. Bach, J. Pankrath, and W. W. Kellogg, eds., *Man's Impact on Climate.* Elsevier Scientific Publishing: Amsterdam.

Myers, N. 1979. *The Sinking Ark.* Pergamon Press: New York.

Myers, P. 1988. The likely impact of climate change on migratory birds in the arctic. Presentation at Seminar on Impact of Climate Change on Wildlife, January 21–22, 1988; Climate Institute, Washington, D.C.

National Research Council (NRC). 1983. *Changing Climate.* National Academy Press: Washington, D.C.

Perring, F. H. 1965. The advance and retreat of the British flora. Pages 51–59 in C. J. Johnson and L. P. Smith, eds., *The Biological Significance of Climatic Changes in Britain.* Academic Press: London.

Peters, R. L., and J. D. Darling. The greenhouse effect and nature reserves. *BioSci.* 35(11): 707–717.

Picton, H. D. 1984. Climate and the prediction of reproduction of three ungulate species. *J. Appl. Ecol.* 21: 869–879.

Rapoport, E. H. 1982. *Areography: Geographical Strategies of Species.* Pergamon Press: New York.

Schneider, S. H., and R. Londer. 1984. *The Coevolution of Climate and Life.* Sierra Club Books: San Francisco.

Seddon, Brian. 1971. *Introduction to Biogeography.* Barnes and Noble: New York.

Strain, B. R., and F. A. Bazzaz. 1983. Terrestrial plant communities. Pages 177–222 in E. R. Lemon, ed., *CO₂ and Plants.* Westview Press: Boulder, CO.

Titus, J. G., T. R. Henderson, and J. M. Teal. 1984. Sea level rise and wetlands loss in the United States. *National Wetlands Newsletter* 6(5): 3–6.

Tralau, H. 1973. Some quaternary plants. Pages 499–503 in A. Hallam, ed., *Atlas of Palaeobiogeography.* Elsevier Scientific Publishing: Amsterdam.

Van Devender, T. R., and W. G. Spaulding. 1979. Development of vegetation and climate in the southwestern United States. *Science* 204: 701–710.

Wigley, T.M.L., P. D. Jones, and P. M. Kelly. 1980. Scenario for a warm, high CO₂ world. *Nature* 283: 17.

Willson, M. F. 1983. *Plant Reproductive Ecology.* John Wiley & Sons: New York.

Chapter 7

DEVELOPING POLICIES FOR RESPONDING TO CLIMATE CHANGE

Jill Jaeger

EXECUTIVE SUMMARY

1. The atmospheric concentrations of a number of trace gases are increasing as a result of human activities. These gases have an important effect in trapping energy at the earth's surface and in the lower

This report is a summary of the discussions and recommendations of the workshops held in Villach, Austria (28 September–2 October 1987), and Bellagio, Italy (9–13 November 1987), under the auspices of the Beijer Institute, Stockholm. It was written by Jill Jaeger (Beijer Institute) based on materials prepared by W. C. Clark, G. T. Goodman, J. Jaeger, M. Oppenheimer, and G. M. Woodwell and contributions from other members of the Steering Committee of the project.

atmosphere (the "greenhouse effect") leading to a warming and thus to changes of climate.

2. It is now generally agreed that if the present trends of greenhouse gas (GHG) emissions continue during the next hundred years, a rise of global mean temperature could occur that is larger than any experienced in human history.

3. A two-stage workshop process held in Villach (Austria) and Bellagio (Italy) in 1987 examined how climatic change resulting from increasing GHG concentrations could affect environment and society during the next century and explored the policy steps that should be considered for implementation in the near term.

4. Scenarios of global climatic change that could occur between now and the end of the next century as a result of continuing emissions of GHGs were developed. The upper-bound scenario, which considers a large increase of GHG emissions and a high sensitivity of the climatic response, gives a global surface temperature increase of 0.8°C per decade from the present until the middle of the next century. The middle scenario, which considers current trends in GHG emissions, a reduction of chlorofluorocarbon emissions according to the Montreal Ozone Protocol, and a moderate climate sensitivity, gives a temperature increase of 0.3°C per decade. The lower-bound scenario, which assumes a strong global effort to reduce GHG emissions and relatively low climate sensitivity, gives a rate of temperature increase of 0.06°C per decade.

5. The most extreme temperature increases would probably occur during winter in the high latitudes of the Northern Hemisphere, where the changes could be two to two and a half times greater and faster than the globally averaged annual values. Precipitation changes could include enhanced winter precipitation in the high latitudes, intensified rains in the presently rainy tropical latitudes, and, perhaps, a decrease in summer rainfall in the mid-latitudes.

6. GHG-induced global warming could accelerate the present sea-level rise, probably giving a rise of about 30 cm but possibly as much as 1.5 m by the middle of the next century. The effects will include: erosion of beaches and coastal margins; land-use changes; wetland loss;

The original publication of this paper provided illustrations that have not been included here. These illustrations depicted global patterns of surface temperature increase as projected by the Goddard Institute for Space Studies model (Hansen et al., 1987), global changes in moisture patterns (Kellogg and Schware, 1981), species-specific latitudinal and altitudinal shifts in range as a result of climate change, and species distribution shifts within biological reserves in response to climate warming.

increased frequency and severity of flooding; damage to port facilities, coastal structures, and water management systems.

7. In the middle latitudes the main impacts are expected to be on relatively unmanaged ecosystems, in particular the forests. If the temperature change is rapid, dieback of trees will result and more and more forest would need managing to maintain it in a productive mode. For the lower-bound scenario of temperature change, extinction of species, reproductive failure, and large-scale forest dieback would not occur before the year 2100. A further effect could be the release of a significant amount of carbon from soils, trees, and other plants as carbon dioxide and methane and this would enhance the greenhouse warming.

8. Climatic change will not occur in isolation. Increasing amounts of atmospheric and aquatic pollutants can be expected from urban-industrial growth. The response to climatic change will be affected by these pollutants. The importance of these interactions and the need to investigate them further cannot be overestimated.

9. In the semi-arid tropical regions the climatic changes that might occur by the middle of the next century as a result of the increasing concentrations of GHGs include a temperature increase of the order of 0.3–5°C and a decrease in precipitation rate in one or more seasons. These changes could worsen the current critical problems of the semi-arid tropics, especially through their effects on food, water, and fuelwood availability, human settlement patterns, and the unmanaged ecosystems.

10. In the humid tropical regions it is expected that the GHG-induced changes could include a warming of 0.3–5°C and an increase in rainfall amount. In addition, tropical storms might extend into regions where they are now less common. Coastal and river regions and regions of infertile soils in the uplands appear to be especially vulnerable.

11. In the high-latitude regions it is expected that the mean winter temperature could increase by between 0.8 to considerably more than 5°C by the middle of the next century. In addition, there could be a withdrawal of the summer pack ice, increased cloudiness and precipitation, slow disappearance of the permafrost, and changes in the tundra and in the northern limit of the boreal forest. These changes can be expected to affect marine transportation, energy development, marine fisheries, agriculture, human settlement, northern ecosystems, carbon emissions, air pollution, and security.

12. The rate of global temperature change that would occur if current trends of GHG emissions were to continue are large compared with observed historic changes and would have major effects on ecosystems and society. For this reason a coordinated international response will become inevitable.

13. Adaptation strategies for responding to a changing climate adjust the environment or our ways of using it to reduce the consequences of a changing climate; limitation strategies control or stop the growth of GHG concentrations and limit the climatic change. A prudent response to climatic change would consider limitation *and* adaptation strategies.

14. Whatever limits on climatic change might be implemented, planning and decision-making could be facilitated by the use of long-term environmental targets, such as the rate of temperature or sea-level change. The choice of a target would be based on observed historic rates of change that did not put stress on the environment or society. The environmental target can be translated into emissions targets for GHGs that could be used for regulatory purposes.

15. An evaluation of the changes of GHG emissions that would be required to limit the global warming rate to the largest natural rates of increase observed in the last century suggests that the limitation could only be accomplished with significant reductions in fossil fuel use.

16. Strategies for adapting to or limiting climatic change could involve high costs to global society. For policymaking purposes there is a need for detailed comparisons of the costs of various strategies.

17. There are many longer-term actions that will be required in order to ensure appropriate responses to climatic changes. The actions that should receive priority now are:

- Approval and implementation of the Montreal Protocol on Substances That Deplete the Ozone Layer.
- Reexamination of long-term energy strategies with the goals of achieving high end-use efficiency; intensification of development of nonfossil energy systems.
- Strong support for measures to reduce deforestation and increase forested area.
- Development and implementation of measures to limit the growth of non-CO_2 GHGs in the atmosphere.
- Identification of areas vulnerable to sea-level rise. Planning for installations near the sea should allow for the risks of sea-level rise.

- Support for and coordination of policy research, global monitoring activities, and policy-directed scientific research on the GHG issue at the national and international levels.
- Examination by organizations, including the intergovernmental mechanism to be constituted by the WMO and UNEP in 1988, of the need for an agreement on a law of the atmosphere as a global commons or the need to move towards a convention along the lines of that developed for ozone.
- Consideration and development of the recommendations of the present report at subsequent conferences, including the World Conference on the Changing Atmosphere (Toronto, June 1988) and the Second World Climate Conference (Spring 1990).

EFFECTS OF CLIMATIC CHANGE ON THE LATITUDINAL REGIONS

OCEANS AND COASTAL AREAS

Half of humanity inhabits coastal regions. The coastal zones are under great pressure due to accelerating population growth, pollution, flooding problems, and upland water diversion.

In these regions the consequences of sea level rise will generally outweigh any direct temperature effects of climatic change, such as enhancement or reduction of fishery resources. In particular, a rise of sea level would, in many places, mean that high tides could penetrate further inland. In addition, the effects of rising sea level would be experienced in terms of greater inland penetration of storm surges.

Global warming induced by greenhouse gases will accelerate the present sea level rise giving a rise of probably about 30 cm and possibly as much as 1.5 m by the middle of the next century, as a result of thermal expansion of the seawater and the melting of land ice. The rate of increase of sea level could therefore be appreciably greater than the 0.01 m per decade long-term average for the last century.

The effects of sea level rise will include:

Erosion of Beaches and Coastal Margins: Some 70 percent of the world's beaches are presently eroding due to a combination of natural sea level rise and human intervention. This process will be aggravated by the sea level rise induced by global warming. For example, the cost of maintaining shores under threat on the East Coast of the USA will be of the order of 10–100 billion US dollars for a one meter sea level rise.

Land-Use Changes: Decrease of usable land for various aspects of primary production, including aquaculture, and for activities such as salt-making arises from the joint effects of subsidence due to river modification and sea level rise. It is generally counteracted by dikes and other diversions. While counteracting measures are not feasible, these changes can lead to large land and livelihood losses. In developed countries, lowland protection against sea level rise will be costly. In developing countries without adequate technical and capital resources, it may be impossible.

Wetlands Loss: Natural wetlands are now under pressure: urbanization encroaches on salt marshes; mangroves are harvested or displaced by fishponds, etc. In the natural state, wetlands adjust to sea level increases by moving landward—a process that is inhibited by steep coastlines and human-made structures. Where migration has become impossible, wetlands will be lost, with associated losses of natural resources, habitat, and physical barriers against flooding.

Frequency and Severity of Flooding: A projected sea level rise of as much as 1.5 m within 75 years would cause many flood disasters, with large losses of life, property, and farmland, particularly in the delta regions of South Asia. Damage to infrastructure in the industrialized countries would also result.

Damage to Coastal Structures and Port Facilities: sea level rise will increase the hydraulic loading on coastal structures like breakwaters, locks, and bridges. Reinforcement will be required and maintenance costs will increase. Port facilities will have to be adjusted to a higher sea level.

Damage to Water Management Systems: sea level rise will also cause problems with drainage and irrigation systems. Saltwater intrusion into groundwater, rivers, bays, and farmland will increase. This will create a demand for reconstruction and extension of water management systems and structures, which may not always be economically feasible. To give an example of costs, in the Netherlands, a country that already has a finely tuned coastal defenses infrastructure, the Public Works Department has tentatively estimated that the minimum adjustments to the water management systems there for a 1-m sea-level rise would require additional investments of the order of several billion dollars.

MID-LATITUDE REGIONS

In the middle latitude regions between 30 and 60 degrees latitude, the amount of warming caused by increasing GHG concentrations will be greater than the global average warming. Additionally, the winter temperatures are expected to increase more than the summer temperatures. Changes in precipitation and soil moisture are uncertain, although soil moisture in summer could decrease as a result of enhanced evapotranspiration with the increased temperatures. The effects of climatic change on agriculture, water resources, and soils have been considered but the most important effects are expected to be on relatively unmanaged ecosystems, especially forests. Two aspects of the global climatic changes are particularly important for these ecosystems:

- Future climatic changes are likely to be much more rapid than those in the past.
- In the absence of measures to limit GHG emissions, the climate will continue to change and the changes will persist indefinitely into the future.

Forests: The forests of the middle latitudes contain, in trees, other plants, and soils, a quantity of carbon that is comparable in magnitude to the amount of carbon currently stored in the atmosphere. The effects of the climatic changes anticipated include the possibility of the release of a significant amount of carbon from this source into the atmosphere as carbon dioxide and methane, which would further increase the greenhouse warming.

The possible responses of mid-latitude forests to climatic change have been assessed, taking into consideration the rate of seed migration, changes in reproductive success with changes in temperature, and climatic stress on standing trees. The reproductive success of many tree species would be reduced by a warming and both tree and plant mortality would increase. These changes would be conspicuous first along the warmer and drier limits of the range of the species. There is no threshold below which effects do not occur.

The upper limit of the climatic changes suggests a rate of warming of 0.8–1.0°C per decade in the middle latitudes. Major effects on forests were estimated to begin around the year 2000 with forest dieback starting between 2000 and 2050. The net effect of the warming would be a reduction in area and standing stock of carbon in forests. The seriousness of the changes depends on the rate of change of temperature.

For the lower bound of the temperature scenario, which gives an average rate of change of mid-latitude temperature of 0.06–0.07°C per decade, extinction of species, reproductive failure, and large-scale forest dieback would not occur before the year 2100, although the warming would cause some changes in the forested regions. However, with the rates of temperature change of the lower scenario the changes in the forested regions would be at rates that are low enough to approximate to the rates at which forests have accommodated to climatic changes observed in the recent past.

These estimates illustrate the importance of the rate of change of temperature for the effects on forests. If the temperature change is rapid, dieback of trees will result, with replacement of successional trees supporting smaller standing crops. The outcome of this trend would be that more and more forest would need planting (or managing) to retain it in a productive mode. Such labor-intensive intervention may be economically nonviable for formerly unmanaged forests.

Agriculture: Warming will cause intraregional shifts in productivity in the mid-latitudes. For all but the most rapid warming, adaptation based on agricultural research should permit maintenance of total global food supplies. However, there will be local disruptions. For the faster rates, agricultural adaptations may be out of step in time with effects of climatic change, generating erratic reductions in food availability. Taken alone, climatic warming would have probably little *net* effect on agriculture in the mid-latitude band: productivity in the lower-latitude zone of the band might be negatively affected because of increased evapotranspiration, while the higher latitudes of the band would benefit from the longer growing season. Agriculture is dependent on the availability of fertile soils. Shifts of crops due to GHG-induced climatic changes may be affected positively or negatively by this factor. There are also major uncertainties about changes in precipitation and evapotranspiration, so that it is not possible to predict at this stage whether the net effects of change will be positive or negative for specific regions except that irrigated agriculture in semi-arid areas in the mid-latitudes will probably be adversely affected by the warming.

Interacting Effects: Climatic change will not occur in isolation. Increasing amounts of atmospheric and aquatic pollutants can be expected from urban-industrial growth. The response to climatic changes will be affected by these increased pollutants. The importance of these interactions and the need to investigate them further cannot be over-emphasized. Such interactions are now causing widespread mortality of many species of coniferous and broad-leaved trees in Europe and in eastern North America.

SEMI-ARID TROPICAL REGIONS

The semi-arid tropical regions lie within the broad latitudinal band 5–35°N and S. Within this zone the semi-arid and subhumid regions are defined as areas receiving mean annual precipitation somewhere within the range 400–1,000 mm, unevenly distributed seasonally, with high spatial and interannual variability.

Climatic variability is a problem for the semi-arid tropical regions. Any future changes in the frequency distribution of extreme events will have important effects. The climatic changes that might occur by the middle of the next century as a result of the increasing concentrations of trace gases in the atmosphere include:

- Increases of regional temperature of the order of 0.3–5°C.
- On average for the latitudinal belt as a whole, climate model results are variable, but a tendency for a decrease in precipitation rate in one or more seasons is generally apparent. In addition, temperature increases would reduce soil moisture availability.

The semi-arid tropical regions already suffer from seasonal and interannual climatic variability. Precipitation data for the zone 5–35°N show a pronounced downward trend since the early 1950s, resulting in prolonged drought and active desertification processes. These regions are very sensitive to climatic variability, generally with negative impacts. Therefore, future climatic changes could worsen the current critical problems of the semi-arid tropics. The major effects are expected to be on:

Food Availability: Temperature increases, precipitation pattern changes, and CO_2 concentration changes would alter the agriculture and agricultural production potential within a region which is already highly sensitive to the impacts of climate and often marginal for agriculture. Productivity changes could aggravate current difficulties in meeting basic nutritional needs. Resource degradation through increased desertification could ensue.

Water Availability: In general, evaporation would increase and runoff would be reduced. Water availability would be further reduced by increased demand.

Fuelwood Availability: Changes in biomass productivity and soil moisture will probably lead to reduced fuelwood availability.

Human Settlement: As agricultural and resource potential changes, human populations are expected to move in response—including increased rural-to-urban migration.

Unmanaged Ecosystems: Climatic changes and human responses are expected to increase pressure upon unmanaged ecosystems and heritage sites. Biotic resources will be stressed by habitat changes and development pressures.

HUMID TROPICAL REGIONS

By the middle of the next century it is expected that the addition of CO_2 and other trace gases will warm the humid tropical regions by 0.3–5°C. This warming, somewhat less than the global average warming, will be accompanied by an increase in rainfall amount, perhaps in the range of 5–20 percent. In a region that is already often too hot and too wet, even such relatively modest climatic changes could have important effects. The increased rainfall may occur largely through increases in rainfall intensity. Superimposed on these general tendencies would be shifts in the geographical patterns of rainfall and cloudiness. Since the warming will increase potential evapotranspiration, there could be a tendency toward more drought stress in many, if not most, of the regions in the humid tropics. With the increased ocean temperatures, tropical storms might extend into regions where they are now less common. Where they already occur, increased intensity of winds and rainfall might be expected.

The major effects of climatic changes would therefore result from:

- Rising water levels along coasts and rivers, resulting from a combination of increasing sea level, greater chance of tropical storm surges, and rising peak runoff. These will result in larger areas being subject to flooding and a risk of salinization.
- Changing spatial and temporal distribution of temperature and precipitation with effects on industry, settlement, agriculture, grazing lands, fisheries, and forests.

Two provinces of the humid tropics appear especially vulnerable to the kinds of climatic change that may occur over the next century:

- Coastal and river regions subject to changes of sea level and storminess.
- Regions of infertile soils in uplands.

HIGH-LATITUDE REGIONS

The high-latitude areas include regions north of 60°N and south of 60°S. The effects discussed here are those that could occur in the northern high latitudes.

The Magnitude of Expected Climatic Changes: By far the largest changes would occur in winter. As a result of increases of GHGs, it is expected that by the middle of the next century the mean winter temperature of this region could increase by 0.8 to considerably more than 5°C. The following effects are the most important:

- Changes of the pack ice conditions could be very great. A warming could result in the withdrawal of summer pack ice, leaving the Arctic ice-free around Spitzbergen and along the north Siberian coast. Loss of pack ice would significantly decrease the proportion of incoming solar radiation that is reflected back to the atmosphere (albedo), which is the reason for the enhancement of the warming effect in these regions.

- There would most likely be increased cloudiness and precipitation in the high-latitude regions of the Northern Hemisphere. Because of the penetration of moisture-rich, warm air into higher latitudes, the precipitation would increase more than evaporation in high latitudes. Thus the rate of runoff into the Arctic Basin would increase markedly. In the 60–70°N region, duration of snowcover would be shorter.
- Permafrost, particularly in northern Canada and Siberia, would slowly disappear.
- Changes in the *tundra and in the northern limit of the boreal forest* will include both a stimulation of growth and carbon fixation and rapid decay of organic matter. The overall effect on carbon storage is not predictable. The possibility exists for a large net release of carbon from soils to the atmosphere as a result of increased respiration.

Changes of Arctic pack ice (including decreased albedo) could have major implications for the climatic changes in lower latitudes. In the absence of the pack ice there would be changes of the atmospheric and oceanic circulations that would cause climatic anomalies in high, middle, and low latitudes. The potential magnitude of these anomalies is not known at present.

Given the above potential changes, the following effects may be expected:

Marine Transportation: The possible changes of sea ice offer opportunities for increased use of the Northeast and Northwest Passages. However, prediction of route enhancement is complicated by inadequate understanding of expected changes of ocean currents, cloudiness, fog, ice fields, and icebergs.

Energy Development: Higher temperatures and a reduction of sea ice could reduce some of the difficulties of offshore oil development, but onshore development could become more difficult and expensive in regions of melting of permafrost, affecting construction practices and existing developments. A reduction of sea-ice extent could lead to higher snowfall over the land surrounding the Arctic Ocean, making operating conditions more difficult.

Marine Fisheries: Different marine ecosystems could be affected positively and negatively by the increased atmospheric and oceanic temperatures. Useful predictions of the effects on fish migration and species distribution will require further research.

Agriculture: Is practiced in Scandinavian countries north of 60°N. At present, however, its importance, as in other circumpolar countries, is small. With warming, agricultural opportunities should improve, but only over limited areas because of lack of suitable soils. Current food-market conditions make it unlikely that extreme northern areas would ever be exploited for the international agricultural markets.

Human Settlement: Climatic warming will make northern mines, forests, and ports more exploitable as growth centers. Opening of the Arctic to shipping will increase the cultural shock which has already stressed native peoples. Warmer climates will induce migration into some areas, putting native peoples at risk of losing traditional cultures and environmental values.

Northern Ecosystems: The changes of precipitation, temperature, and sea level will affect the natural ecosystems. Rapid shifts in growth conditions could cause dislocation or disruption of ecosystems as well as movement of the limits of agriculture and forestry northwards.

Carbon Emissions: Nordic regions are important in the global carbon cycle. It has been suggested that the climatic warming could result in a substantial increase of methane emissions from tundra, thus increasing the emissions of GHGs into the atmosphere. Siberian and other boreal soils are often highly organic and would rapidly decay upon withdrawal of permafrost, if they subsequently dry up, thus increasing CO_2 loading to the atmosphere.

Air Pollution and Acid Rain: Arctic haze (already circumpolar) and acid deposition in nordic regions would be affected by climatic changes. The result could be a shift of the region subject to acid deposition, particularly if atmospheric circulation patterns change. There could be a significant degradation of some aquatic and terrestrial ecosystems and perhaps an improvement of others. The importance of these interactions and the need to investigate them further cannot be overemphasized.

Security: The northern ice-bound land borders of North America, Europe, and Siberia are highly sensitive national defense zones for all states with Arctic territory. If these coastlines become navigable, fundamental security readjustments will be required.

PART III

GLOBAL WARMING: PHYSICAL IMPACTS

Chapter 8

GREENHOUSE EFFECTS AND IMPACTS ON PHYSICAL SYSTEMS

John Firor

The scientific study of climate heating by infrared-absorbing gases in the atmosphere—the so-called greenhouse effect—has been a century-long attempt to attach the proper numbers to a relatively simple physical process. The radiative balances within the atmosphere are strongly affected by the existence of small concentrations of gases in the air, most notably carbon dioxide, that absorb infrared radiation and hence complicate the flow of heat from the earth's surface to space. It is a simple calculation to show that if infrared-trapping gases could be removed from the air and all else remained the same, the earth's surface would be 33°C colder than now—a very large cooling that would result in an ice-covered earth. The expectation that the climate will heat in coming decades arises naturally from this same simple calculation: if the historical concentration of infrared-absorbing gases in the atmosphere heats the earth's surface by such a large amount, will not the increases in concentration that result from human activities heat the climate even more?

The obvious answer to this question is yes. But a *useful* answer should also include the rate of heating at each location on earth, given a projection of changes in heat-trapping gases in the atmosphere. An even more useful answer would also estimate the changes in precipitation, soil moisture, sea level, weather variability, cloudiness, and winds that would accompany the heating. Producing these useful estimates is a gigantic task which is the province of the numerical modelers of the atmosphere, working with those who can document specific physical processes in the climate system, and aided by those who can measure and reconstruct present and past climates. This work has attracted numerous experts who have labored quietly over many years and who have contributed to periodic reports by government agencies and scientific academies on the progress of their effort to simulate, with useful precision, the workings of the global climate. The following five chapters recount their overall progress and present some of the details.

Chapter 9 summarizes the scientific basis for the greenhouse effect. Early work on climate and climate change depended on one-dimensional simulations of the climate which allowed scientists to work out details of the transfer of radiation through the atmosphere and which suggested the two and three-dimensional effects that needed to be incorporated into later calculations. As larger computers were developed, the models assumed more dimensions, more complexity, and a greater number of feedbacks. Today, a few research groups around the world have constructed three-dimensional, global circulation models with realistic continents, dynamics, and radiative transfer, with approximate treatment of clouds and surface hydrology, and with some representations of the ocean surface, sea ice and ice caps, and other phenomena. These groups also have access to sufficiently large computers that calculations covering tens or even hundreds of simulated years can be accomplished and a variety of experiments can be completed in which the computer output is compared with actual climates or in which climate heating is simulated.

THE STANDARD EXPERIMENT: ITS USES
AND ITS PROBLEMS

One aspect of this process needs to be noted, since it has led to misunderstandings outside the modeling community. It has been the custom of the climate modeling groups to perform a standard experiment in climate heating. This experiment consists of simulating the climate for a number of years using an atmosphere with a "low" amount of carbon dioxide, usually taken to be 300 ppm by volume, and then repeating the simulation with a doubled (600 ppmv) concentration of carbon dioxide. In each simulation, the calculation is carried forward long enough for the model to come to equilibrium. The climate for that particular computer run is then derived from an average of the conditions for years at the end of the run when equilibrium has been established. The difference between the two calculations can then be represented, if needed, as a single number: the global average temperature increase to be attributed to the doubled carbon dioxide.

The reason for adopting this standard is that it makes possible the easy comparison of the results with the results of earlier calculations by the same group or with the model results of other investigators. Climate heating calculations are most frequently reported in terms of this experiment, and models are sometimes described in terms of their "sensitivity"—that is, how much average temperature increase is shown in the standard experiment. In the reporting of experiments designed to determine the heating that will be forced by increases in the heat-trapping gases other than carbon dioxide, therefore, one reads: "Using a model with a sensitivity of 2.7°C, we find . . ." National and international groups convened to review the progress of climate change science frequently couch their reports in terms of this experiment, as well. Thus one of the most widely known facts relating to this issue is that the consensus value for climate heating is between 1.5 and 4.5°C. It should be noted, however, that recent studies with general circulation models show sensitivities of over 4°C, near the upper end of this range.

In recent years it has been realized that there are two problems connected with the standard experiment. The first is that the concentrations of infrared-trapping gases are in fact increasing yearly, rather than making one large jump. The behavior of the climate forced by this continuous change is likely to be quite different from the climate described by the standard calculation. In particular, the amount of heat taken up by the oceans will differ markedly from the amount absorbed by the land surfaces. This difference, combined with the fact that the oceans are far from uniformly spread over the earth, suggests that the heating will appear at the earth's surface in a complex and changing pattern quite unlike the pattern that would exist at equilibrium.

The second problem is that reporting on the standard experiment tends to leave two mistaken impressions with the nonscientific reader, whether policymaker or interested citizen. One impression is that carbon dioxide is the only gas involved in climate heating, whereas it is estimated that other greenhouse gases, though appearing in the air in quite small concentrations, together have a heat-trapping effect at least as great as that of carbon dioxide alone. The most important gases that contribute to this rapid heating are methane, the two common chlorofluorocarbons (CFC-11 and CFC-12), nitrous oxide, and tropospheric ozone. There is also a long list of other gases, mostly industrial chemicals, each of which makes a slight addition to the heat trapping. Methane is emitted from a variety of sources where organic material decays in contact with a limited oxygen supply—such as swamps, rice fields, municipal waste dumps, the digestive tracts of cattle and termites—and as a result of fossil fuel production. Other pollutants resulting from fossil fuel burning interfere with the reactions which remove methane from the atmosphere and thus increase its atmospheric concentration. The CFCs are synthetic chemicals produced for a number of uses; nitrous oxide comes from fuel burning and from using nitrogen fertilizers. Ozone is not emitted directly to the lower atmosphere but is created in reactions driven by sunlight that start with nitrogen oxides and other atmospheric pollutants. It is likely that considerably more than half the total greenhouse warming results directly from fossil fuel use; most of the rest comes from deforestation and the CFCs.

The other troublesome impression created by reports of the standard experiment is that the climate is visualized as heating by 3 or 4°C and then stabilizing. Hence the policy discussion sometimes focuses on what kind of readjustments will be needed to live in a stable but hotter world. There is, in fact, no natural process foreseen that will level off emissions of heat-trapping gases at any particular value—control of these gases depends on policy decisions yet to be made. As a consequence, policy needs to focus on how we should respond to a rapidly changing climate and at what point we must control climate change. These false impressions may diminish in coming years as modeling groups gain access to computers large enough to deal with continuously changing amounts of infrared trapping and as review groups begin to report their consensus conclusions in terms of rates of climate heating or rates of sea-level rise rather than as a fixed value for an arbitrary increase in one gas. These improvements are already under way.

When climate specialists meet policy experts, another problem appears—one that is perhaps more psychological than related to physical science. In the last two decades, as the powerful social implications of climate heating have become widely recognized, a widespread feeling has developed that the current temperature of the earth's surface must somehow be a special one held in place by negative-feedback mechanisms and able to withstand unintended attempts by the human race to change it. This feeling has given rise to many ideas about how such feedbacks might work, and these notions have received considerable publicity at meetings and in the press. Thus while builders of complex numerical simulations have been striving to determine the sensitivity of their models to the inclusion of certain physical processes—and trying, too, to convince government funding sources of their need for large computers that would allow higher spatial resolution in the model and a greater degree of coupling the model atmosphere with models of oceans, ice, and land surfaces—much of the scientific community and the public at large have been hearing about the latest process that will cancel any additional climate heating or ameliorate one of its major effects.

The unfortunate aspect of this situation is that the burden of proof—the job of working out all the details and making a convincing demonstration that the proposed feedback is quantitatively significant—is apparently lifted from the proposers of the new process (perhaps because we all hope that what they suggest is true) and the modelers are looked upon as behind the times since they have not yet taken full account of the "newly discovered" effect. Suggestions of positive feedbacks, which would *accelerate* the climate heating, are also common. These proposals tend to receive less publicity, but often they too need more careful review by their proposers to ascertain whether they are sufficiently large to affect the long-range climate projections. Despite these distractions the modelers have, step by step, incorporated the major physical processes acting on and within the climate, they have tested their models against actual climates, and they have produced numbers of increasing credibility.

CLIMATE PROJECTIONS: A QUESTION OF RELIABILITY

The construction of a climate model is an intricate and fascinating story, but for the policymaker the focus is on another question: Just how reliable are these climate projections? After scientists have incorporated into the models as many of the relevant variables as they can fit into the world's largest computers, after they have calculated what the climate will be like with more heat-trapping gases in the air, and after they have reported their findings carefully and in cooperation with other scientists, the public and the policymaker are still left with the problem of deciding how much trust one can place in model calculations that seem to foresee, as these do, a dramatic and possibly overwhelming climate change.

To resolve this difficulty one must examine how the models have been used. Climate models have received intense scrutiny in recent times and have been tested in a number of ways. The large climate models have been used, for example, to calculate today's climate, and the results have been compared in great detail with what we actually observe. The models have been used to compute past climates where reconstructions are available and, again, comparisons made. They have been used to calculate the difference between summer and winter—a brief but dramatic climate change. And related models have been used to compute the climate of Mars, which has much less of a heat-trapping atmosphere than we do, and the climate of Venus, which has much more.

One could give more examples, but the point is this: there has been a vigorous and continuing effort to examine the usefulness of climate models. In this examination the models have passed the increasingly difficult tests to which they have been subjected, and confidence in their results has grown. There is today a strong scientific consensus that the global heating projected by these models must be taken seriously.

There are still, however, areas of uncertainty and controversy. Just how rapid will the average effect be? Different models give different estimates. Using a particular scenario of heat-trapping gas emission, for example, the consensus global heating rate is about 0.4°C per decade. Using the same emissions scenario, one model gives a rate as low as 0.2 while another gives 0.7. This wide range points to the different structure and emphasis built into each model, as well as the caution that arises from the realization that some of the important interactions in the atmosphere are represented quite crudely in the models. Moreover, one can never prove that no important process has been omitted from the models—something crucial that would change the answer in a major way—so the modelers continue to be cautious in their claims for numerical accuracy. The gradual building of confidence through a variety of tests, and the continuous observation and interpretation of interactions in the climate system, are the only known ways of guarding against overlooked but significant processes. This wide range of possible heating rates has less effect on the policy discussions than might be thought, however, since even the smallest values represent a faster shift of climate than has ever been experienced by human civilizations and therefore require a thoughtful policy response.

A second area of uncertainty is in local effects. How much of the heating will show up at each location on earth? Will each place get more or less rain? There is a strong consensus among scientists that the picture of the average heating is about right, but there is still much work to do to get the details straight. All the models show the world getting warmer, for example, but not all of them show the midwestern United States getting drier. Chapter 10 discusses progress in resolving this uncertainty. But a comprehensive ability to estimate local manifestations of climate heating will depend not only on the availability of the larger computers which will permit the simulation of more detail, but also on an improved understanding of those features of ocean circulations that modify sea surface temperature and hence the interaction of the atmosphere and the ocean. One regional impact of special interest is the change that will be produced in hydrology by the climate heating. Soil moisture and stream flows will both be influenced by any change in the total amount or seasonal distribution of precipitation and by changes in evaporation induced by changes in temperature. Chapter 11 discusses major changes in water supplies in the western United States that might accompany a global heating.

A third uncertainty, discussed more fully in Part II of this volume, concerns emission feedbacks. In addition to the physical feedbacks within the climate system itself, such as the change in cloudiness that may accompany a change in temperature, there are concerns that the emissions of both carbon dioxide and methane may increase with rising temperature. An increase in global temperature will thaw some polar regions containing peat, and the subsequent decay of the peat could add to the atmospheric burden of carbon dioxide. Globally, a temperature rise could increase the respiration rates of forest ecosystems and enhance carbon dioxide emissions. Methane increases could arise from enhanced activity of soil microorganisms or from the release of methane complexes now sequestered at the bottom of shallow, cold seas. As these positive-feedback mechanisms could have a great effect on global heating, each of these processes needs to be studied in detail in order to quantify its role in climate change.

The timing and rate of the heating contain uncertainties associated with the role of the oceans in the climate. Some of the additional heat at the earth's surface will be absorbed by the ocean rather than appearing as an increase in surface air temperature. This process will delay the surface heating by an interval that is estimated to be a few decades, and some such delay is included in the climate change scenarios announced by review groups. But the true length of the delay depends on the sensitivity of the atmosphere to increasing concentrations of infrared-absorbing gases and on facts about ocean circulations that are still not well understood.

The heating and other changes will first become apparent to us as extreme climatic events become more frequent. The probability of a hot year, for example, or an unusual string of days in which the temperature exceeds a certain value each day will increase noticeably with even a small change of average temperature. One effect of the heating will be truly global: the rise of sea level as the higher temperatures expand the ocean water and add to it by melting glaciers and ice caps. People living along the Atlantic and Gulf coasts of the United States well understand that a hurricane's damage to beaches and property depends on whether it hits the coast at high tide or low tide. With the increase in sea level, all storms will come ashore on higher water than they did previously, and all damage will be increased. Land that is now not far above sea level will have to be defended with dikes or returned to the sea. These changes are discussed in some detail in Chapter 12.

Finally, there is one uncertainty that represents an additional threat. Given that society's main difficulty with climate change arises from its rapidity, anything that increases that rate also increases the strains placed on people and institutions. Many scientists who have spent careers studying the workings of the earth–ice–atmosphere–ocean–biosphere system are impressed with the observation that a slowly increasing force on some part of the system can sometimes produce a sudden and rapid adjustment. From the sudden overturning of the waters in a lake to the surging of a glacier, from the collapse of a fishery to the change in the course of a river, the world is full of examples of slow buildups of pressures followed by sudden adjustments of the system. The "ozone hole" that grew rapidly over the Antarctic in response to a gradually increasing concentration of chlorine-bearing gases in the stratosphere illustrates this possibility, as do certain climate events reconstructed from evidence left by the ice age glaciations. Modelers think they are unlikely to be able to foresee such events—the possibilities are too numerous and the conditions leading to a dramatic shift too special, as they were over the Antarctic. This additional uncertainty, described in Chapter 13, suggests that the projections of heating, sea level rise, and shifts in other physical features of the climate produced by models may represent a less damaging, more gradual scenario than will in fact occur.

LESSONS OF THE VILLACH CONFERENCE

A recent review of the progress in studying all these issues took place in Villach, Austria, in September and October 1987. This workshop, attended by scientists from many countries, produced a summary report presenting the current consensus on the rates of climate change. Assuming a continuation of current emission trends, the group estimated that the global temperature would heat at about 0.3°C per decade and the sea would rise at 5.5 centimeters per decade. This report (see Chapter 7) also attempted to give as much regional detail as the studies warrant.

The Villach conferees considered the impact of these climate changes region by region and heard a variety of expert opinion on the difficulty or impossibility of adaptation to such rapid changes. In the mid-latitudes, natural forests could not migrate rapidly enough to move with changing climate, so forests could not be maintained without huge labor-intensive efforts. Coastal engineers explained that dikes could be built to hold back the sea in critical places, but the world does not have enough money to pay for it. The high-latitude experts noted that it would be easier to run ships in the Arctic as floating ice melted, but fragile arctic ecosystems could be devastated, indigenous cultures would be disrupted, and the permafrost would thaw—exposing vast peat beds to decay, releasing great quantities of carbon dioxide and methane to the air, and thus speeding up the whole process. These presentations shifted the focus of the meeting to consideration of what we can do now to slow the warming and thus have a reasonable chance to adapt.

At this point a new meaning to "human impact" appears, one we have not directly faced before. If we allow ever-increasing changes in the composition of the air, and the accompanying major climate change, we will have forced ourselves to replace working ecosystems with activities of our own design. The world's ecosystems have evolved over long ages along with the ocean and atmosphere. And since we have little evidence that we know enough to replace all the services these systems supply to each other and to us, this is a risky course indeed. In the first place, we may not be smart enough to pull it off. And even if we do, we will have abandoned other options—there is no turning back once the evolved systems are gone. And we have not even considered whether this kind of world—entirely managed by people—is the kind of world we want.

Eventually the composition of the atmosphere must be stabilized, and it would seem wise to set a goal of stabilization at a level that does not force dramatic global consequences. The exact numbers are not yet known, but this goal clearly requires that emissions into the atmosphere be set at rates that allow the air to cleanse itself—and such rates are well below the current emissions of each of these gases. Such a goal requires greatly improved energy efficiency and a rapid transition to nonfossil energy. It probably also means that we must not destroy all the forests, that we must reduce our total industrial activity as we know it today, and that we must face certain troublesome population issues. Such a goal— to stabilize the composition of the air so that the world we live in bears some relationship to the one that evolved over millions of years—will be a very difficult one indeed. But the alternative is to let the composition of the air change indefinitely, and accept the utterly artificial world that would ensue.

Chapter 9

SCIENTIFIC BASIS FOR THE GREENHOUSE EFFECT

Gordon MacDonald

INTRODUCTION

Compared with the intense heat of Venus, or the stark day–night contrast of Mars and the moon, the earth possesses a relatively equitable climate that makes possible the existence of Homo sapiens and all their activities. The favorable nature of the earth's climate is made possible by the beneficial interactions of the oceans, atmosphere, and biosphere. The definition of climate is complex but a primary descriptor

Testimony given in a joint hearing before the Subcommittees on Environmental Protection and Hazardous Wastes and Toxic Substances of the Committee on Environment and Public Works, U.S. Senate, One-hundredth Congress, first session, 28 January 1987.

The original publication of this paper provided illustrations that have not been included here. These illustrations depicted the estimated variation of atmospheric methane concentration (Stauffer et al., 1985) and the trend in mean annual ozone concentration at Rügen, German Democratic Republic (after Warmbt, 1979).

of climate is temperature. The temperature of the earth's atmosphere depends on the delicate balance between the incoming, short-wavelength radiation of the hot sun and the outgoing radiation of the much cooler earth.

If the atmosphere were stripped away, the average temperature of the earth's surface would be an inhospitable 35°C colder than it is today. The atmosphere provides a warming blanket for the surface through the action of certain constituents whose concentration is much smaller than that of the dominant nitrogen and oxygen. Water vapor, carbon dioxide (CO_2), and ozone (O_3) are the most important of these species, though a myriad of other constituents contribute to the warming.

The oceans play an important role in determining climate not only by providing a gigantic reservoir for water (and carbon dioxide) but also through their large thermal inertia. Without the stabilizing oceans, our climate would be much more subject to rapid change. The biosphere also serves as a source and sink for climate-controlling constituents of the atmosphere. By determining in part the reflective nature of the earth's surface, the biosphere also controls, in part, that fraction of the sun's radiation that is reflected back into space without interacting thermally with the surface and atmosphere.

Weather and climate are important determinants of a region's character and how a society develops. The day-to-day fluctuations of the atmosphere make up weather. By convention, *normal* climate is taken as a moving 30-year average of weather. The 30-year standard is an attempt to balance the practical limitation of data accumulation with statistical requirements for stable measures. Despite the artificiality of such definitions, it is clear that weather and climate influence a region's energy use, its growth of plants and animals, its means of transportation, its water supplies, its natural areas, its pattern of habitation; in brief, its infrastructure. A combination of temperature and availability of moisture determines whether vegetation is forest, prairie, tundra, or desert, and also the ease with which society can adapt to natural stresses. The frequency of such extreme events as tornadoes, hurricanes, floods, and droughts affects the habitability of a region and the investments required to moderate economic, physical, and social stresses created by weather and climate. The entire physical and economic infrastructure of a region is an adaptation to the prevailing climate, an adaptation developed over centuries.

Climate change refers to shifts in normal climate lasting over decades. Year-to-year events, such as several years of extremely dry or wet conditions, are variations in climate, not climate change, if they are within the expected statistical deviations for normal climate. Decades-long trends in regional climate, such as the gradual change of a tropical savanna to a desert, constitute climate change. Global climate change refers to shifts, generally in the same direction, over all parts of the planet. The Little Ice Age, dating from about 1500–1850 A.D., involved a global cooling with an extension of sea ice in both the Northern and Southern Hemispheres. The Little Ice Age thus qualifies as climate change.

Global climate changes have had a profound effect on human history (Lamb, 1982). Only 18,000 years ago, a several-kilometer-thick ice sheet covered much of Canada, northern United States, and northern Europe. During the period 8000–4000 years before the present (B.P.), conditions were warmer than today by perhaps as much as 1–2°C, with precipitation zones shifted northward. About 1000 years B.P., relatively warm temperatures allowed the Norse to cross the North Atlantic. With the Little Ice Age came temperatures a few tenths of a degree centigrade colder than today. Not only were temperatures lower but there was an increased variability in climate with numerous years of famine and plague well documented in Europe.

We now recognize that man's activities can bring about changes in climate in addition to those due to the natural fluctuations and long-term evolution of the atmosphere. Before nineteenth-century industrialization, man's activities were of too small a scale and too low an intensity to alter weather or climate. With the mechanization of agriculture and the greatly enhanced use of carbon-based fuels, particularly coal, the situation changed. Both the burning of coal and the greater development of agriculture began the release of carbon that had been stored in the soil and rocks for thousands or millions of years. Once released, carbon forms atmospheric carbon dioxide and the composition of the atmosphere is changed, perhaps irreversibly. Of all of man's myriad activities, none has brought about as pervasive and large-scale an alteration as that of the composition of the atmosphere resulting from agriculture and energy use.

Though Tyndale (1861) recognized that slight changes in atmospheric composition could bring about climate variation, it was not until

1938 that Callendar showed that man, by burning carbon-based fuels, was capable of changing the climate of the planet. The subject of anthropogenic climate change is thus only 50 years old. Intensive study began with the heightened interest in the environment during the late 1960s. The last decade has seen two major advances in understanding climate change. Carbon dioxide, once thought to be the sole culprit, is not alone in changing atmospheric thermal balance; other gases, taken together, are of comparable importance. Stated in another way, looking toward the future, the problem is twice as severe. The second discovery is that the earth's surface in the 1980s is 0.5°C warmer than it was during the 1880s. The change has not gone smoothly over the years, but despite the erratic course of the earth's temperature, it is clearly moving in the direction of global warming in accord with theoretical expectations.

THE ATMOSPHERE AS A GREENHOUSE

The hot surface of the sun radiates as a body having the equivalent temperature of 5800°K. The bulk of the radiation is in the visible-wavelength region, 0.4–1.0 μm, where the earth's atmospheric gases absorb only weakly. In contrast, the low-temperature earth emits radiation at infrared wavelengths for which the atmosphere is highly absorbing, as is illustrated in Figure 9.1. In an oversimplified description, the atmosphere lets shorter-wavelength radiation in but does not let the longer-wavelength radiation out. By supposed analogy with the behavior of panes of glass, the effect is called the *greenhouse effect*. In reality, greenhouses do not depend on trapping radiation for their warmth but rather inhibit cooling of the plants due to moving air; thus, thin plastic works just about as well as glass.

In the strongly absorbing infrared region, different molecular species are responsible for the opaqueness of the atmosphere at various wavelengths (see Figure 9.1). Water molecules are strong absorbers over much of the region at which the earth radiates, 5–30 μm. However, water is a relatively weak absorber in the 8.5–18 μm interval, which coincides with the region in which the intensity of the earth's outgoing radiation is at a maximum (see Figure 9.1). The 12.5–18 μm region is blocked by the carbon dioxide now in the atmosphere. The addition of

FIGURE 9.1

Absorption of the Earth's Thermal Emissions

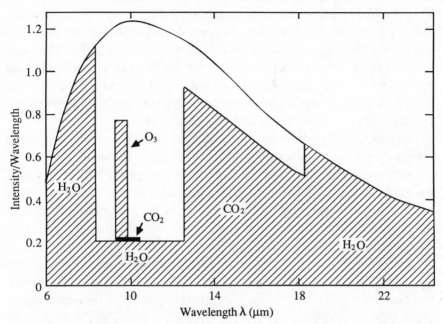

Wavelength λ (μm)

Thermal emissions of the earth's surface at a temperature of 290 K. The amount of emissions absorbed, and the relative values of absorptive gases, are indicated by shaded areas. The Figure illustrates why trace amounts of absorbers in the 8.5 to 12.5 μm (micron) region can be critically important in determining the thermal balance of the atmosphere.

further carbon dioxide will not affect the 12.5–18 μm region of outgoing radiation. The 8.5–12.5 μm interval is relatively transparent so that energy radiated at these wavelengths flows more freely into space. However, transparency of the 8.5–12.5 μm "window" is reduced in part by strong absorption by ozone at 9.6 μm and weaker absorption by CO_2, as shown in Figure 9.1. Increasing concentrations of ozone and carbon dioxide thus reduce the transparency of the window leading to a warming of the atmosphere. In addition to molecular species, clouds and particulate matter (aerosols) can contribute to the trapping.

Calculating the combined radiation-trapping effect of a variety of molecular species is a complex task. The contributions of individual absorbers do not add linearly, in part because molecules differ greatly in their capacity to absorb radiation and also because the regions which the species block off overlap. In rough terms, CO_2 currently accounts for 12% of the trapped radiation, but if all other absorbers were removed, CO_2 would trap three times as much energy. Similarly, clouds currently trap 14% of the radiation, but they would trap 50% of the currently absorbed radiation if all other absorbers were removed. The complexities of the radiative processes are further underlined by noting that the upper layers of the atmosphere leak relatively more radiation into space than they trap, so that additional carbon dioxide leads to atmospheric cooling rather than warming at layers above 20 km.

Over the last decade, it has been recognized that a variety of species have strong infrared absorption modes that lie in the window of transparency (Ramanathan, 1975; Chamberlain et al., 1982). In some cases, the concentrations required to affect the radiative balance are minute compared with that of carbon dioxide, due to the strength of the absorption bands. Trace species that are capable of affecting the radiative balance include nitrous oxide (N_2O), methane (CH_4), and freons (CCl_3F, CCl_2F_2), as well as ozone. All of these trace species are increasing in concentration due to agricultural and industrial activity, and these increases add to the blanketing capacity of the atmosphere.

CHANGES IN ATMOSPHERIC CARBON DIOXIDE CONCENTRATION IN HISTORICAL TIMES

Remarkably, carbon dioxide was the first constituent of the earth's atmosphere to be identified. Joseph Black published the discovery during 1754 in a University of Edinburgh thesis. By 1870, quantitative measurements developed which compare in precision with today's measurements. Interpretation of early observations is made difficult not by inadequacies in analytical chemistry but because they were not undertaken to measure *changes* in the composition of the atmosphere. Sampling for this purpose is complicated owing to the extreme danger of contamination from, for example, the observer's breath. While numerous observations of the CO_2 content of the atmosphere are reported in the literature during the period 1870–1910, their interpretation remains ambiguous. After about 1910, few reports are available since the composition of the atmosphere was considered known and fixed and analytical chemists pursued other interests.

The situation changed dramatically in 1958 when C. D. Keeling of the Scripps Institution of Oceanography, at the suggestion of Roger Revelle, began a series of painstaking measurements of CO_2 concentration on a remote site at what is now the Mauna Loa Observatory in Hawaii. His observations, continued to the present, show an exponential growth of atmospheric carbon dioxide on which a periodic seasonal fluctuation is superimposed (Keeling et al., 1984). (See Figure 5.1 in Chapter 5.) The seasonal variation follows the biospheric uptake of CO_2 during the growing season and its discharge during winter. Keeling also established and maintained an observing regimen at the South Pole. The Antarctic observations show a similar exponential growth but a more diminished seasonal variation. Since the atmosphere at the South Pole is well removed from biological activity, the smaller seasonal fluctuations are expected. Keeling's observations have been duplicated at other stations in various parts of the world over shorter periods of time. In all sets of observations, the exponential increase is clear, though the amplitude of the seasonal terms varies with latitude. For example, a 20-year-long record taken from observations on a weather ship in the northeast Pacific (50°N, 145°W) shows greater seasonal fluctuations than those seen at Mauna Loa; the amplitude of these fluctuations reflects the strong seasonal dependence of the mid-latitude Northern Hemisphere biological activity.

The exponential growth in atmospheric CO_2 concentration correlates with the release of CO_2 by the burning of oil, coal, gas, and wood. Since 1945, the injection of carbon into the atmosphere by fuel combustion appears to be exponential, with a slowing in growth following the 1973 explosion in energy prices (see Figure 9.2). Exponential fits to the Mauna Loa and South Pole bracket the increase in carbon-based fuel use with the following exponential doubling times:

Mauna Loa	26.6 years
Fuels	23.9 years
South Pole	21.9 years

While the growth in atmospheric CO_2 parallels the release of carbon through energy use, the overall carbon budget is not well understood. The observed increase in atmospheric CO_2 accounts for only about half the carbon released through energy and industrial uses. Oceans are generally assumed to be a major sink for CO_2. While numerous models have been constructed to mimic ocean-wide diffusive and convective

FIGURE 9.2
Anthropogenic CO$_2$ Production

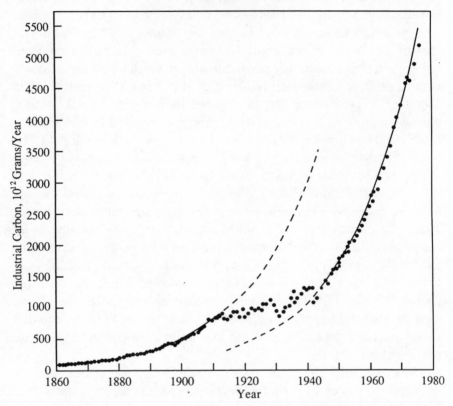

The rate of production of carbon dioxide by man's activities (after Keeling, 1984). Units are 10^{12} g yr^{-1}. The major deviations from exponential growth between 1914 and 1945 reflect the effects of the two world wars and the worldwide economic chaos between the wars.

exchange processes, the ocean's capacity to remove CO$_2$ from the atmosphere over times on the order of decades remains largely uncertain. The carbon balance mystery is deepened by inadequate data on the net contribution of biospheric processes to the atmospheric carbon burden. Destruction of forest cover and cultivation of new land provides a source of atmospheric CO$_2$, but stronger growth in the many plants stimulated by increased CO$_2$ concentration should lead to greater biological productivity, and thus lead the biosphere to serve as a sink for CO$_2$. Both the sources and sinks associated with the biosphere are large numbers so that the net impact, being the difference of these large numbers, is most uncertain.

Estimates of future atmospheric carbon dioxide levels are clouded by questions regarding the overall carbon budget. A rough guess is provided by the observation that historically about half the fossil fuel carbon placed in the atmosphere remains there. More refined estimates will require a better definition of the role of the oceans and biosphere in the carbon cycle. Since oxygen is also removed from the atmosphere during the formation of CO_2, long-term measurements of the oxygen content of the atmosphere may be required to sort out the various sinks and sources for carbon.

The concentration of CO_2 in the atmosphere was about 315 ppm by volume when Keeling began his observations in 1958. During the period 1958–1986, CO_2 concentration increased by about 30 ppm, or 9.5% in 28 years. Central to understanding the total impact of intensive agriculture and energy use on the present CO_2 concentration is to establish the concentration 100 years ago, when these activities were just beginning. There are a number of ways to obtain a preindustrial value. The first is to assume that conditions prevailing over the last 30 years held during the preceding 70. If one-half the carbon released in combustion remains in the atmosphere and all other sources and sinks are relatively constant, then the concentration of CO_2 in the atmosphere during 1890 was about 293 ppm. Two ways of checking this value are available—analysis of air trapped in glaciers and a careful reanalysis of the older observations.

When snow is transformed into ice during glacier growth, air is trapped in the intergrain spaces. The air remains in the airtight glacial ice as the glacier continues to grow. Chemical analysis of this air provides an estimate of the CO_2 content of the atmosphere at the time the air was encased. A number of difficulties lessen the value of such determinations. The time at which a particular layer of ice was laid down is difficult to estimate. The rate of snow accumulation affects the quantity of air that is trapped. The seasonal melting of snow may incorporate new air, leading to confusion. Despite these uncertainties, the values obtained from analysis of the most favorable samples indicate an atmospheric CO_2 concentration in 1890 of between 280 and 295 ppm (Raynaud and Barrola, 1985). These values are in general agreement with the value obtained by extrapolating current CO_2 concentration backward, assuming constant relative sources and sinks of CO_2.

From and Keeling (1986) have carried out a detailed reexamination of early observations of CO_2 concentrations, particularly those made by Callendar, the discoverer of the significance of anthropogenic CO_2. Among the criteria From and Keeling used was whether early observations showed a seasonal cycle in accord with modern observations. The result of their analysis is that the mean annual concentration of atmospheric CO_2 circa 1880, in uncontaminated air at 50°N latitude, was 292 ppm. This value is consistent with backward extrapolation of current values as well as estimates derived from glacial air samples.

Historical observations coupled with extremely careful modern measurement show that the CO_2 concentration of the atmosphere has increased from about 290 ppm to 345 ppm in 100 years, or 19%. Future concentrations of CO_2 are difficult to estimate since they depend on unknown energy use rates, mixtures of fuels, patterns of deforestation and agriculture, and uptake of CO_2 by the oceans and atmosphere. Uncertainties in future CO_2 levels due to uncertainties in the carbon cycle may be small if past behavior provides a reliable clue; if this is the case, then about one-half of the CO_2 produced by fossil fuel combustion ends up in the atmosphere. There remains the difficult task of projecting energy use decades into the future if we are to determine the atmosphere of tomorrow and its associated climate.

TRENDS IN OTHER WARMING GASES

Historical trends of other anthropogenically produced gases, including methane, nitrous oxides, and ozone, are poorly known, in part because recognition of their significance to climate change has come only in the last decade. Freons present a different story; they are manufactured gases and came into major use only during the 1960s. Production data are readily available, except for Eastern Bloc countries and China, and since freons are very long-lived in the atmosphere, production data alone provide a good guide to their trends since introduction during the 1960s.

Anthropogenic sources of methane include direct emissions from the production and transportation of natural gas and from the burning of biomass. Cultivation of rice paddies and the raising of ruminant animals can also lead to increased methane production. Interpretation of the observed methane concentration is complicated in that a number of atmospheric chemical reactions involving methane and other constituents, such as carbon monoxide, can lead to increases or decreases in atmospheric methane concentration. Further, emissions of carbon monoxide are increasing as the use of fossil fuels increases. Observations made over the past decade at a number of stations show that the CH_4 concentration is increasing at about 2% per year (Khalil and Rasmussen, 1983, 1985). Whether the increase is due to greater emissions or heightened removal of constituents in the atmosphere that would otherwise react with methane is not clear. As in the case of carbon dioxide, data on past methane concentrations can be obtained from the analysis of trapped air in glaciers. These data suggest a long-term exponential growth with an exponential doubling time of 46 years. Given the large uncertainties in interpretation of the air bubble data, the close fit of observation with the exponential fit should be viewed with some skepticism, particularly since the sources and sinks for methane are not well understood.

While the overall residence time of methane in the atmosphere is relatively short, about 6 years, nitrous oxide is much less reactive and has a residence time of over 100 years. The current annual rate of increase of N_2O is estimated at 0.2% per year (Weiss, 1981). This value is based on sparse observations; historical data are not available. A number of processes add N_2O to the atmosphere. Combustion of fuel produces N_2O, as does the breakdown of ammonia used as a fertilizer. Disposal of human and animal waste strongly stimulate nitrification, and waste disposal is another source of atmospheric N_2O.

Freons came into major use during the 1960s and initially showed a high, 10–15% per year growth. Due both to the imposition of environmentally based regulation and to the downturn in economic growth, the rate of freon use slowed during the late 1970s and early 1980s, but has resumed more recently (Cunnold et al., 1983a, 1983b). Most recent data indicate a growth rate for $CFCl_3$ of 9.0 parts per trillion (ppt) per year and for CF_2Cl_2 of 15.3 ppt per year (Cunnold et al., 1986). These values correspond to percentage growth rates for both $CFCl_3$ and CF_2Cl_2 of 5.1% per year. The observed atmospheric concentrations and growth rates are in good agreement with published production data (Cunnold et al., 1986).

A variety of observations show that the tropospheric concentration of ozone is increasing, even though the stratospheric concentration may be decreasing as a result of complex chemical reactions with freons stimulated by ultraviolet radiation. While the total amount of ozone in the troposphere is small compared with stratospheric ozone, it is important in considering the energy balance of the atmosphere. At the lower levels of the atmosphere, ozone is at higher pressures and its absorption of outgoing infrared radiation is very much more effective than higher-level ozone. Measurements from a rural site on an island in the Baltic show an increase in ozone of about 60% since the 1950s (see Fig. 4 in Warmbt, 1979). Similar trends have been observed at stations in the Alps, North America, and Japan, though local atmospheric conditions complicate interpretation of the data (Logan, 1985).

As in the case of methane and nitrous oxide, the sources and sinks of ozone are not thoroughly understood. Ozone forms photochemically in urban smog but can also form in rural atmospheres if hydrocarbons are present. Free hydroxyl radical (HO) is a principal sink for ozone, but HO may be taken up by methane and carbon monoxide with which it also reacts. Increasing concentrations of CH_4 and CO will lead to increased tropospheric ozone. A principal source of carbon monoxide is the automobile. Increased automobile traffic leads to increased levels of hydrocarbons and nitrogen oxides—sources of ozone and of carbon monoxide which is a sink for hydroxyl. The combination of higher source concentrations and lower sink (hydroxyl) concentration leads to higher ozone levels and increased atmospheric warming.

While the importance of methane, freons, nitrous oxide, and ozone in producing a greenhouse effect is well established, their past and future concentrations are much less certain. All are the result, either directly or indirectly, of energy use and/or industrial activity. Thus, all of these radiatively active species can be expected to increase significantly over the next 50 years.

MODELS OF CLIMATE CHANGE DUE TO GREENHOUSE GASES

The task of predicting future climate change is extremely complex. Given the relative stability of climate evidenced by the observed slow rate of climate change over geologic time, it is tempting to seek a past climate that would provide an analog to present and future climates. Increasingly detailed data are available on conditions prevailing during the Holocene climatic optimum (5000–6000 years B.P.), the last major interglacial period (about 125,000 years B.P.), and the Pliocene warming (3–4 million years B.P.). During these intervals, the mean temperature was warmer than today by about 1°C, 2°C, and 3–4°C. But these times are poor analogs to future climatic change. Past changes took place slowly, on time scales of at least thousands of years; their cause is not known. Today, changes in the composition of the atmosphere are taking place in decades, the basic radiative characteristics of present constituents are nonlinear, and changes in composition are taking place at an exponential rate. While past climates may provide guidance as to the possible future evolution of warm periods, they yield less information as to current trends. An alternative to extrapolating from historical climates is to construct artificial ones from mathematical models.

Over the years, a hierarchy of climate models has been developed to estimate climate change. At one end of the scale are energy balance models which are based on conservation of energy and, in their simplest form, treat the entire earth's surface as having a single temperature. Their advantage is their simplicity; the only equation that enters is that of conservation of energy, unknown quantities need not be guessed at, and the calculation does not require a computer. The disadvantage of such energy balance models is that only average temperature changes are predicted; this failing must be weighed against the advantages of parsimony of assumptions and simplicity of calculations. At the other end of the scale of complexity are the Global Circulation Models (GCM). These are three-dimensional representations of the atmosphere that explicitly include equations for conservation of mass, water vapor, energy, and zonal and meridional components of momentum. In their most detailed form, GCMs are based on a 500-km grid size in the numerical approximation to horizontal motions, with up to nine layers in the vertical direction. As a result, on the order of 20,000 grid points are required in approximating the continuous atmosphere over the earth's surface. In total, the calculation requires the treatment of about 200,000 equations, with each equation requiring extensive calculations. The calculation of the evolution, over time, of this large set of equations strains the largest available computers, particularly when careful attention is paid to delicate issues of numerical stability of the finite difference approximation to the underlying partial differential equations.

Comprehensive three-dimensional models are designed to provide estimates of those parameters of climate change most critical to society, the regional, seasonal, and diurnal distribution of temperature, soil moisture, and surface runoff. This task is made complicated by the requirement that valid approximations be developed for the many processes taking place over the time and space scales of interest. For example, there is a need to represent clouds and cloud systems which typically have dimensions of 1 to 10 km, much less than the resolving power of the numerical grid of 500 km. Major difficulties arise in parameterizing cloud coverage and cloud heights because these quantities vary on scales less than the numerical resolution. Because clouds are important in calculating the distribution of radiative energy, the manner in which the approximations are made is critical, and various climate modeling groups use somewhat different approaches. Similar difficulties in parameterization arise in treating soil moisture, air–sea interaction, topography, biological activity, etc.

Among models of complexity intermediate to the energy balance and GCMs are the radiative-convective models (RCM) (Ramanathan and Coakley, 1978). These treat the vertical structure of a global average atmosphere assuming buoyantly stable stratification. RCMs divide the vertical atmosphere into several layers but do not deal with horizontal motions. RCMs are particularly well suited to a detailed evaluation of the radiative effects of CO_2 and trace gases and aerosols, as well as the chemical reactions among radiatively active species.

A conventional means of comparing models is to calculate the average surface temperature that would result from doubling the atmospheric CO_2 concentration from 330 ppm to 660 ppm. Early estimates made during the late 1960s predicted that a doubled CO_2 content should raise the average temperature 1.5–3°C (Manabe and Wetherald, 1967). More than 100 independent estimates of average surface temperature increase have since been made using energy balance, RCMs, and GCMs. Almost all of these estimates lie in the range 1.5–4.5°C, with values near 3°C tending to be favored (Schlesinger and Mitchell, 1985).

Intercomparisons of GCM simulations of CO_2-induced climate change show that while there is general agreement as to the surface temperature change, there is much less agreement in the geographical distribution of temperature changes, and sometimes large disagreement on geographical distribution of soil moisture and runoff. This state of affairs is not surprising in that the calculation of temperature depends primarily on energy conservation, and all models incorporate this basic feature. Geographic distributions of various quantities, particularly those involving precipitation, will depend sensitively on the details of the parameterization of those quantities that are poorly understood or inadequately resolved in the numerical approximations.

The large heat-retaining capacity of the oceans coupled with the mixing processes in their upper layers are the main factors providing thermal inertia to the climate system. This thermal inertia slows down climate change due to anthropogenic changes in atmospheric composition. GCMs can be used to calculate the delay of the equilibrium response of the atmosphere to changes in atmospheric composition. Results from available models suggest that thermal storage and mixing in the ocean reservoirs cause a lag in the warming of 10 to 20 years. If an exponential change of atmospheric CO_2 composition with a doubling time of 20 years leads to an equilibrium warming of 1°C, then the observed response would only be 0.7 ± 0.2°C (Hoffert and Flannery, 1985). The large uncertainty arises because of incomplete knowledge of the formation of bottom waters and of the mixing processes in the wind-stirred upper layers of the oceans, particularly the Arctic and Antarctic pumps involving thermohaline circulations.

Over the past 100 years, the carbon dioxide content has increased by about 20%. A range of model calculations suggests that the corresponding equilibrium temperature rise should be 0.5–1.0°C. If this is corrected for the effects of the thermal inertia of the oceans, the changing composition of the atmosphere should have produced a warming of 0.35–0.7°C superimposed on the natural fluctuations of the atmosphere. Since natural variations associated with El Niño and volcanic eruptions bring about short-term changes on the order of a few tenths of a degree, the roughly half-degree increase of the longer-term trend must be extracted by careful statistical analysis.

OBSERVED CLIMATE CHANGES

In addition to noting the anthropogenic warming, Callendar (1938, 1949, 1961) attempted to estimate expected changes in climate, particularly temperature shifts. As interest in the subject has increased, larger computer-based data sets have been developed. The principal problem in analyzing the data is one of sampling. Data are highly biased toward land in the Northern Hemisphere. Since oceans cover more than 70% of the earth's surface, this bias is large. Many land-based observing stations are located near urban sites. This location itself introduces a further bias since as the urban region grows, more energy is consumed and there is a local warming trend. Inconsistencies in data collection stem from changes in both observing instruments and calibration practices. Observing stations have been relocated, and the shelter used to house the thermometer may have been altered. These are just a few of the difficulties that must be overcome in order to construct a homogeneous data set. While all of these considerations are important, the principal deficiency in the past has been the lack of data over the oceans.

Jones et al. (1982, 1986a) have carried out a detailed analysis of land-based observations. The data have been carefully examined to detect and correct for nonclimate errors that result from station shifts, urban development, and changes in instrumentation. Folland et al. (1984) and Jones et al. (1986b) have attempted the daunting task of using sea surface temperatures (SST) and marine air temperatures (MAT) recorded by ships at sea to obtain ocean temperature estimates. For MAT, there are problems, including a great diversity of instruments and practices as well as varying placement of the thermometers on ships. SST are determined both from bucket measurements and from water intake measurements, after the conversion from sail to steam. Jones et al. (1986b) attempt to calibrate the SST and MAT observations by first carrying out a correlation of MAT with nearby land for 15 regions of the world. The corrected MAT for various time periods are then used to calibrate the SST. The number of observations is large; Folland et al. (1984) used a British data set of 46 million SST observations. Jones et al. (1986b) used that set and the Comprehensive Ocean Atmosphere Data Set of the National Oceanic and Atmospheric Administration comprising some 63 million SST observations.

The results of the analysis of Jones et al. (1986b) are summarized in Figure 9.3. In judging the results shown, it should be recalled that spatial coverage is at best 75% and, more importantly, the coverage changes with time. Coverage before 1900 is less than 30%, with the

FIGURE 9.3
Annual Mean Temperature Variations Since 1861 (Jones et al., 1986b)

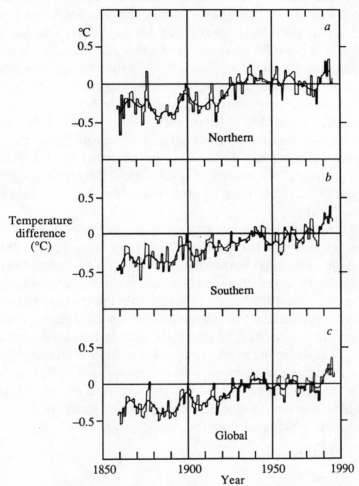

Global change portrayed in curve c, *Northern Hemisphere in curve* a, *and Southern Hemisphere in curve* b.

Northern Hemisphere always much better represented than the Southern Hemisphere. In particular, the ocean south of 45°S and the whole southeastern Pacific are underrepresented.

The curve in Figure 9.3 shows a long-term warming trend. The magnitude of the warming agrees with theory; it is about 0.5°C over the last 100 years. In view of the heightened interest in long-term climatic change, the question naturally arises as to whether this trend would have been noticed if theory had not predicted that it should be there. I was convinced of the warming long before the detailed analysis of the records were available. I had observed that the snouts of the glaciers on the Alps on the South Island of New Zealand had moved from sea level to high up the mountain between 1900 and the present. New Zealand, having a relatively isolated geographical setting in the ocean, was more likely to capture longer-term trends in climate than glaciers in more continental regions. But glaciers, even on an isolated island, may not well serve as a surrogate for overall climate change. Since theoretical predictions of a one-half degree warming preceded the statistical analysis of historical observation, critics undoubtedly will question the reality of the derived warming even though the statistical base for the inference is strong.

The record of the warming trend shown in Figure 9.3 has several interesting features. The three warmest years were 1980, 1981, and 1983. Five of the nine warmest years in the 134-year record occurred after 1978. However, there remain puzzles. For the 40-year period between the mid-1930s and the mid-1970s the average temperature remained approximately stable at a time the exponential change in atmospheric composition should have produced an increased rate of warming. The increase in temperature between 1905 and about 1930 is high when compared to the expected increase. Thus, while the overall change in temperature is in the right direction and of the correct magnitude, a number of features of the observed warming are inconsistent with those expected from the greenhouse effect.

FUTURE GREENHOUSE WARMING

Projections of future warming are uncertain since future levels of use of carbon-based fuels are not known. The contemporary budgets of methane and nitrous oxides are not known, so projections of the contribution of these gases to future climatic shifts are even more uncertain than that of CO_2. However, over a relatively short interval of 50 years, 1980 to 2030, limits can be placed on expected climate variations on the assumption that there will be no radical shifts in the world's agricultural-

FIGURE 9.4
Changes in Concentration of Radioactive Gases for the Period 1980–2030

Species	Residence Time (Years)	1980 Mixing Ratio (ppbv)	Estimated 2030 Mixing Ratio (ppbv)
CO_2	2	340×10^3	450×10^3
N_2O	120	300	350–450
CH_4	7–8	1650	2500–3000
CCl_2F_2	110	0.30	0.6–1.2
CCl_3F	65	0.18	0.5–1.0
O_3 (troposphere)	0.1–0.3	20–70	30–100
CO	0.3	90	100–200

industrial base. Ramanathan et al. (1985), using a radiative-convective model, have calculated an equilibrium warming of 1.5°C between 1980 and 2030, assuming a stable projection of current energy use. A little less than half of the projected increase is due to CO_2; the remaining increase is due to the other radiatively active gases. If the thermal inertia estimate of about 0.7 of the theoretical warming holds, then the expected warming during the next 50 years is about 1°C, or twice that shown for the interval 1880–1980. The total warming over the 150-year period ending in 2030 of 1.5°C would be equal to or larger than that estimated as taking place during the Holocene, 5,000 to 6,000 years ago.

Using the simpler energy budget model, I have calculated a comparable temperature change (MacDonald, 1985). The assumed increases in the warming gases are shown in Figure 9.4. Also included is carbon monoxide, which is an infrared absorber but whose concentration influences the abundance of methane and ozone. The calculated equilibrium surface warming is 1.8–2.2°C, as indicated in Figure 9.5. This corre-

FIGURE 9.5
Surface Warming Due to Change in Atmospheric Composition: 1980–2030

Component	ΔT_s in °K
CO_2	1.0
O_3 (troposphere)	0.2–0.4
CH_4	0.3–0.4
N_2O	0.2
$CFCl_3$ and CF_2Cl_2	0.1–0.2
Total	1.8–2.2

sponds to an actual warming of 1.3–1.5°C, if current estimates of the thermal lag of the oceans are appropriate. The contribution to the warming of species other than CO_2 is 80–120% of the CO_2 warming, with ozone and methane being the principal other contributors. The major difference between these estimates and those of Ramanathan et al., other than the model basis, is that they anticipate a significantly greater contribution of freons to the warming (0.44°C) and lower contributions from ozone and methane. The lower contribution from ozone is due to Ramanathan et al.'s estimate of only a 12.5% growth in 50 years, which seems low when compared with recent trends.

While the magnitudes of future sources and sinks of radiatively active species other than CO_2 are uncertain, it appears that the contribution of these gases to future warming can be as great or greater than that due to CO_2. The average global temperature in the year 2030 should be 1–1.5°C warmer than today. The climate in this regime will differ markedly from that we have become accustomed to.

CONCLUSIONS

Despite the many uncertainties in future energy use, carbon cycle evolution, and production of trace gases, it appears highly probable that over the next 50 years anthropogenic activities will warm the earth's surface by 1–1.5°C. This warming will be superimposed on the half-degree warming of the last 100 years. These changes are larger than those that can be anticipated from natural variations and are of the same order or larger than those associated with the Holocene warming 6,000 years ago. If GCMs and previous climates are any guide, it is likely that the effects will be noted first at high latitudes. There will be a tendency toward lower precipitation and soil moisture in the mid-continent regions of North America and Eurasia, with more frequent, extended droughts. The warming may be accompanied by much increased climatic variability, with heightened frequency of extremes.

Much current thinking about the greenhouse effect assumes that society will adapt incrementally to change, even if the changes involve heightened sea level and droughts in the "food basket" areas of North America and Eurasia. These assumptions are not convincing. Infrastructures of society such as water supplies, transportation nets, and land use patterns have evolved over centuries in response to prevailing climate. Significant changes in climate over decades will exert profound disruptive forces on the balance of infrastructures. The prevailing market economies in much of the world adjust to alterations in supply and demand through adjustments in price. The intangibles of climate are not readily quantified in conventional market terms; they are one of the externalities with which markets find so hard to deal.

A move away from carbon fuel-based economies, whether stimulated by market forces or government intervention, provides one alternative, if predictable climate changes become unacceptable. A candidate energy technology is nuclear, but this alternative poses far greater fears to much of the world's population than does climate change. Whether such fears are justified is irrelevant; the perception of dangers from the operation of nuclear power plants and storage of waste is real.

Conservative alternatives to that of blind belief in incremental adjustment are preparation of the way for alternative energy sources, including nuclear and renewable sources, as well as increasing overall energy efficiency. While climate change may be acceptable in the short term, it is not clear that it will be in the longer term.

References

Callendar, G. 1938. The artificial production of carbon dioxide and its influence on temperature. *Q.J. Roy. Meteor. Soc. 64*, 223–237.

Callendar, G. 1949. Can carbon dioxide influence climate? *Weather 4*, 310–314.

Callendar, G. 1961. Temperature fluctuation and trends over the earth. *Q.J. Roy. Meteor. Soc. 87*, 1–12.

Chamberlain, J., H. Foley, G. MacDonald, and M. Ruderman. 1982. Climate effects of minor atmospheric constituents. In *Carbon Dioxide Review: 1982*, ed. by W. Clarke. Oxford University Press, 255–277.

Cunnold, D., R. Prinn, R. Rasmussen, P. Simmonds, F. Alyea, C. Cardelino, A. Crawford, P. Fraser, and R. Rosen. 1983a. The atmospheric lifetime experiment: 3, Lifetime methodology and application to 3 years of $CFCl_3$ data. *J. Geophys. Res. 88*, 8379–8400.

Cunnold, D., R. Prinn, R. Rasmussen, P. Simmonds, F. Alyea, C. Cardelino, and A. Crawford. 1983b. The atmospheric lifetime experiment: 4, Results for CF_2Cl_2 based on 3 years of data. *J. Geophys. Res. 88*, 8401–8414.

Cunnold, D., R. Prinn, R. Rasmussen, P. Simmonds, F. Alyea, C. Cardelino, A. Crawford, P. Fraser, and R. Rosen. 1986. Atmospheric lifetime and annual release estimates for $CFCl_3$ and CF_2Cl_2 from 5 years of ALE data: 3. *J. Geophys. Res. 91*, 10797–10817.

Folland, C., D. Parker, and F. Kates. 1984. Worldwide marine temperature fluctuation 1856–1981. *Nature 310*, 670–673.

From, E., and C. Keeling. 1986. Reassessment of late 19th century atmospheric carbon dioxide variation in the air of Western Europe and the British Isles based on unpublished analysis of contemporary air masses by C. S. Callendar. *Tellus 38B*, 87–105.

Hoffert, M., and B. Flannery. 1985. Model projections of the time-dependent response to increasing carbon dioxide. In *The Potential Climate Effects of Increasing Carbon Dioxide*, ed. by M. MacCracken and F. Luther. U.S. Dept. of Energy, DOE/ER-0237, Washington, D.C.

Jones, P., T. Wigley, and P. Kelly. 1982. Variations in surface air temperatures: Part 1. Northern Hemisphere 1881–1980. *Mon. Weather Rev. 110*, 59–70.

Jones, P., S. Raper, R. Bradley, H. Diaz, P. Kelly, and T. Wigley. 1986a. Northern hemisphere surface air temperature variations 1851–1984. *J. Climate Appl. Meteor. 25*, 161–179.

Jones, P., T. Wigley, and P. Wright. 1986b. Global temperature variation between 1861 and 1984. *Nature 322*, 430–434.

Keeling, C. 1984. *Atmospheric CO_2 Concentration, Mauna Loa Observatory, Hawaii, 1958–1983*. U.S. Dept. of Energy Report NDP-001. Carbon Dioxide Information Center, Oak Ridge, Tenn.

Keeling, C., T. Wharf, C. Wong, and R. Bellagag. 1985. The concentration of atmospheric carbon dioxide at Ocean Weather Station P from 1969 to 1981. *J. Geophys. Res. 90*, 10511–10528.

Khalil, M., and R. Rasmussen. 1983. Sources, sinks, and seasonal cycles of atmospheric methane. *J. Geophys. Res. 88*, 5131–5141.

Khalil, M., and R. Rasmussen. 1985. Causes of increasing atmospheric methane: Depletion of hydroxyl radical and the rise of emissions. *Atmos. Environ. 19*, 397–407.

Lamb, H. 1982. *Climate, History, and the Modern World*. Methuen, London.

Logan, J. 1985. Tropospheric ozone: Seasonal behaviour, trend and anthropogenic influence. *J. Geophys. Res. 90*, 10463–10482.

MacDonald, G. 1985. *Climate Change and Acid Rain*. The MITRE Corporation MP8600010, McLean, VA.

Manabe, S., and R. Wetherald. 1967. Thermal equilibrium of the atmosphere with a given distribution of relative humidity. *J. Atmos. Sci. 24*, 241–259.

Ramanathan, V. 1975. Greenhouse effect due to chlorofluorocarbons: Climate implications. *Science 190*, 50–51.

Ramanathan, V. and C. Coakley. 1978. Climate modeling through radiative-convective models. *Rev. Geophys. Space Phys. 16*, 465–489.

Ramanathan, V., R. Cicerone, H. Singh, and J. Kiehl. 1985. Trace gas trends and their potential role in climate change. *J. Geophys. Res. 90*, 5547–5566.

Raynaud, D., and J. Barrola. 1985. An Antarctic ice core reveals atmospheric CO_2 variations over the past few centuries. *Nature 315*, 309–311.

Schlesinger, M., and J. Mitchell. 1985. Model prediction of the equilibrium climatic response to increased carbon dioxide. In *The Potential Climate Effects of Increasing Carbon Dioxide,* ed. by M. MacCracken and F. Luther. U.S. Dept. of Energy, DOE/ER-0237, Washington, D.C.

Stauffer, B., G. Fischer, A. Nefter, and H. Oeschinger. 1985. Increase of atmospheric methane recorded in Antarctic ice core. *Science 229,* 1386–1388.

Tyndale, J. 1861. *Philadelphia Magazine, Journal of Science 22,* 169–194, 273–285.

Warmbt, W. 1979. Results of long-term measurement of near surface ozone in the G.D.R. *Z. Meteoral. 29,* 24–31.

Weiss, R. 1981. The temporal and spatial distribution of tropospheric nitrous oxide. *J. Geophys. Res. 86,* 7185–7196.

Chapter 10

CHANGES IN SOIL MOISTURE

Syukuro Manabe

I would like to focus my presentation on the large-scale changes in soil wetness. These changes, which may have profound agricultural and other implications, have received increased emphasis in research activities at various institutions in North America and Europe. I will discuss this issue based upon the results (references 1 and 2) from the mathematical models of climate developed at the Geophysical Fluid Dynamics Laboratory of NOAA.

The mathematical model of climate used for this study is a three-dimensional model of the atmosphere coupled with models of the land surface and a simple mixed-layer ocean. The atmospheric part of this climate model is very similar to the numerical weather prediction models which have been used successfully by the National Weather Service of the United States and many other countries for daily weather forecasting. The model takes into consideration the effects of solar and terrestrial radiation and the hydrologic cycle and explicitly follows the general circulation in the atmosphere by the hydrodynamical equations.

Testimony given in hearings before the Committee on Energy and Natural Resources, U.S. Senate, One-hundredth Congress, first session, on the Greenhouse Effect and Global Climate Change, 9–10 November 1987.

It has been shown that this type of climate model successfully simulates the seasonal and geographical distribution of climate. The impact of increased greenhouse gases on climate was evaluated by comparing the two climates of the model with normal and above-normal concentrations of atmospheric carbon dioxide.

A map of the CO_2-induced percentage change of soil moisture for the June–July–August period is illustrated in Figure 10.1. This figure indicates that, in summer, soil becomes drier over very extensive, mid-continental regions of North America, southern Europe, and Siberia in response to the doubling of atmospheric carbon dioxide. In some regions, the reduction amounts to a substantial fraction of the soil moisture present in the standard-CO_2 case.

Over Siberia and Canada, CO_2-induced changes in snowmelt are responsible for the reduction in soil moisture. In these regions extensive snow cover prevails during winter before melting in the late spring. Since snow cover reflects a large fraction of incoming sunshine, its disappearance increases the absorption of solar energy by the land surface to be used as the latent heat for evaporation. Thus, the end of the spring snowmelt season marks the beginning of the seasonal drying of the soil that takes place in summer. In the warmer high-CO_2 world the snowmelt season comes earlier, bringing an earlier start of the spring to summer reduction of soil moisture. As a result, the soil becomes drier in summer.

Over the Great Plains of North America the earlier snowmelt season also contributes to the CO_2-induced reduction of soil wetness in summer. Changes in the mid-latitude precipitation pattern also contribute to the reduction of soil wetness in summer over North America and southern Europe. Both of these regions are under the influence of a rainbelt associated with the typical path taken by mid-latitude low-pressure systems. In the high-CO_2 atmosphere, warm moisture-rich air penetrates further north than in the normal-CO_2 atmosphere. Thus, the precipitation rate increases significantly in the northern half of the mid-latitude rainbelt, whereas it decreases in the southern half. Since the rainbelt moves northward from winter to summer a mid-latitude location is in the northern half of the rainbelt in winter and in its southern half in summer. At such a location the CO_2-induced change in precipitation rate becomes negative in early summer, contributing to a reduction of soil moisture. The summer dryness is enhanced further due to the

FIGURE 10.1

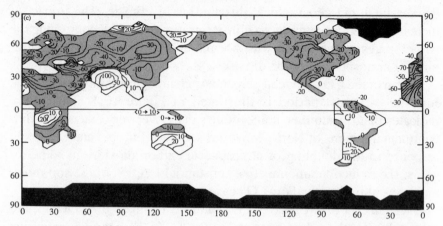

The geographical distribution of the percentage change of soil moisture in response to a doubling of CO_2 content for the June–July–August period. The results are obtained from a general circulation model of climate developed at the Geophysical Fluid Dynamics Laboratory of NOAA.

increased sunshine reaching the ground as reduced evaporation from the drier continental surface causes a decrease in cloudiness.

The summer reduction of soil wetness discussed above does not continue through winter. In response to the increase of atmospheric carbon dioxide, soil wetness increases over extensive mid-continental regions of middle and high latitudes in winter. In middle latitudes, this is mainly due to the increase of precipitation in the northern half of the middle-latitude rainbelt. In high latitudes, a larger fraction of precipitation is realized as rainfall, making soil wetter.

Upon inspecting Figure 10.1, one should keep in mind that only these very broad-scale features of the soil moisture changes are significant. For example, many of the small-scale features in the tropics and Southern Hemisphere are not regarded with much confidence. This is partly because the climate models used in this study have coarse computational resolution and fail to simulate the small-scale features of the hydrologic change. Furthermore, the detailed features of the CO_2-induced change are often obscured by the natural fluctuation of soil wetness, thereby making the identification of these features very difficult. This large natural hydrologic variability also implies that it will take much longer to detect the mid-continental enhancement of summer dryness than the global warming of the atmosphere.

One should also note that research groups have not reached unanimous agreement (references 3 and 4) on the issue of the mid-continental summer dryness, though the results from recent studies (references 5 and 6) appear to agree with those presented here. In my opinion, some uncertainty in the estimate of the future hydrologic change stems from the difficulty of reliably incorporating into a climate model various relevant physical processes, such as the land-surface water budget and ocean–atmosphere interaction. Nevertheless, the above discussion clearly indicates that the reduction of the mid-continental soil moisture in summer results from global warming of climate and is a rather robust and large-scale phenomenon. In conclusion, I believe it is likely that mid-continental summer dryness will intensify in the Northern Hemisphere accompanied by the increasing greenhouse warming of climate.

From the above discussion, it is essential that increased research effort be devoted to the improvement of climate models to increase our confidence in their projection of CO_2-induced climate change. In addition, the results presented here could provide an added incentive for improved water management. By the same reasoning research for the development of agricultural species which can survive under significantly different environmental conditions is encouraged. Progress in both of these areas would better prepare us for the natural fluctuations of climate as well as climate change due to greenhouse gases.

References

1. Manabe, S., and R. T. Wetherald. 1986. Reduction in summer soil wetness induced by an increase in atmospheric carbon dioxide. *Science 232,* 626–632 (2 May 1986).
2. Manabe, S., and R. T. Wetherald. 1987. Large-scale changes of soil wetness induced by an increase in atmospheric carbon dioxide. *J. of Atmos. Sci. 44,* 1211–1235.
3. Schlesinger, M. E., and J.F.B. Mitchell. 1985. Model projection of equilibrium climate response to increased CO44255. In MacCracken, M. C., and Luther, F. M. (eds.), *Projecting the Climatic Effects of Increasing Carbon Dioxide.* DOE/ER-0237. U.S. Department of Energy, Washington, D.C. Pp. 81–147.

4. MacCracken, M. C., M. E. Schlesinger, M. R. Riches, and S. Manabe. 1986. Atmospheric carbon dioxide and summer soil wetness. A letter and a response. *Science 234,* 659–660 (7 November 1986).
5. Mitchell, J.F.B. 1986. *Dynamical Climatology.* Tech. Note 39. Meteorological Office, Brackness, Berkshire, England.
6. Schlesinger, M. E. 1987. Personal communication.

Chapter 11

EFFECTS OF CLIMATIC CHANGE ON WATER SUPPLIES IN THE WESTERN UNITED STATES

Roger R. Revelle and Paul E. Waggoner

In this chapter we show that warmer air temperatures and a slight decrease in precipitation would probably severely reduce both the quantity and the quality of water resources in the western United States. Similar effects can be expected in many water-short regions elsewhere in the world. We have not attempted to estimate these, primarily because we do not know enough to be able to do so. But we hope that hydrologists of other countries will be stimulated by our calculations to investigate the probable consequences of a CO_2-induced climate

This chapter originally appeared in *Changing Climate,* copyright 1983, by the National Academy of Sciences. (The section containing the details of the analysis summarized in Table 7.4 is not included.)

change on water resources in their own countries. In all countries, planning and construction of large-scale water-resource systems take many decades. The time involved is of the same order of magnitude as the time over which a significant change in climate from increase of carbon dioxide and other greenhouse gases can be expected. Thus, we believe that planners and managers of water systems throughout the world should be able to make good use of forecasts of the hydrologic consequences of a warmer climate and of possible changes in precipitation.

EMPIRICAL RELATIONSHIPS AMONG PRECIPITATION, TEMPERATURE, AND STREAM RUNOFF

To assess the effects on the United States' water resources of probable climatic change we used the empirical relationship found by Langbein et al. (1949) among mean annual precipitation, temperature, and runoff. This was based on representative data from 22 drainage basins in the conterminous United States. Their relation in Figure 11.1 gives the estimated annual runoff for different values of mean annual precipitations and weighted mean annual temperatures. The latter were computed for each catchment basin by dividing the sum of the products of average monthly temperature and precipitation by the mean annual

FIGURE 11.1
Runoff (mm yr^{-1}) as a Function of Precipitation and Temperature

Weighted Average Temperature (°C)	Precipitation (mm yr^{-1})					
	200	300	400	500	600	700
− 2	54	92	154	230	330	440
0	40	74	124	190	275	380
2	28	57	95	154	225	320
4	17	40	78	125	190	265
6	9	25	60	100	155	220
8	0	17	42	82	128	185
10		8	29	64	103	155
12		0	19	47	80	130
14			10	32	65	105
16			0	20	50	85

(*Source: Langbein et al., 1949.*)

precipitation. In this way, the average temperature during each month is weighted by the precipitation during that month.

The catchments studied by Langbein and his colleagues were distributed over climates from warm to cold and from humid to arid, but in Figure 11.1 we have shown the relations among runoff, temperature, and precipitation only for relatively arid areas. In these arid areas, the value of actual evapotranspiration is less than the potential evapotranspiration that would occur if sufficient water were present and evapotranspiration was controlled mainly by temperature. For example, at a temperature of 4°C, potential evapotranspiration is about 450 mm yr^{-1}. Yet Langbein's data show that even when annual precipitation is only 300 mm, there is still significant runoff, about 13% of the precipitation. Correspondingly, average actual evapotranspiration must be only 260 mm yr^{-1}.

From Figure 11.1 we observe that for any given annual precipitation, runoff diminishes rapidly with increasing temperature. Similarly, for any given temperature, the proportion of runoff to precipitation increases rapidly with increasing precipitation. For example, at a weighted mean annual temperature of 4°C and annual precipitation of 200 mm, runoff is only 8.5% of precipitation, whereas for the same temperature and an annual precipitation of 700 mm, runoff is 38% of precipitation. At an annual temperature of 8°C, runoff is zero when precipitation is 200 mm or less and is 185 mm—or 26.4% of precipitation—when the average annual precipitation is 700 mm.

For any particular region, the relations shown in Figure 11.1 are rather crude approximations because many physical factors, including geology, topography, size of drainage basin, and vegetation, may alter the effect of climate on runoff. We believe, nevertheless, that these relationships can be used without serious error to describe the effects of relatively small changes in average temperature and precipitation on mean annual runoff.

In Figure 11.2, we have used the data in Figure 11.1 to compute the approximate percentage decrease in runoff for a 2°C increase in temperature. Climate models (e.g., Manabe and Wetherald, 1980) indicate that a temperature change of this magnitude or greater is likely as a result of the doubling of carbon dioxide and increased concentration of "greenhouse gases" expected during the next century. We see that for a present weighted mean annual temperature of 4°C and annual precipitation of 300 mm, a 35% diminution in runoff would follow a 2°C

FIGURE 11.2

Approximate Percentage Decrease in Runoff for a 2°C Increase in Temperature

Initial Temperature (°C)	Precipitation (mm yr⁻¹)					
	200	300	400	500	600	700
− 2	26	20	19	17	17	14
0	30	23	23	19	17	16
2	39	30	24	19	17	16
4	47	35	25	20	17	16
6	100	35	30	21	17	16
8		53	31	22	20	16
10		100	34	22	22	16
12			47	32	22	19
14			100	38	23	19

(*Computed from* FIGURE 11.1.)

warming. The percentage decrease in runoff from a warming diminishes with increasing precipitation and becomes greater for successively higher values of the initial temperature.

Figure 11.3 shows the approximate percentage decreases in runoff for a 10% decrease in precipitation. According to the results of climate models, such a diminution in precipitation is likely, at least in certain regions of the United States, with a doubling of atmospheric CO_2. Again we see that the effect becomes larger with higher average annual

FIGURE 11.3

Approximate Percentage Decrease in Runoff for a 10% Decrease in Precipitation

Temperature (°C)	Initial Precipitation (mm yr⁻¹)				
	300	400	500	600	700
− 2	12	16	17	18	18
0	14	16	17	19	19
2	15	16	19	19	20
4	17	19	19	21	21
6	23	23	21	21	21
8	30	24	24	22	22
10		24	27	23	23
12		40	30	25	25
14			34	30	27
16			50	36	29

(*Computed from* FIGURE 11.1.)

temperatures. There is a relatively small difference in the percentage decrease in runoff at a given temperature over the range of initial precipitation values shown in the table. Comparison of Figures 11.2 and 11.3 shows that below an initial mean annual precipitation of 500 mm, the effects of a 2°C warming are larger than those caused by a 10% decrease in precipitation. The reverse is true when mean annual precipitation is 500 mm or more.

EFFECTS OF CLIMATE CHANGE IN SEVEN WESTERN U.S. WATER REGIONS

Stockton and Boggess (1979) have used Langbein's empirical relation to estimate the effects of a climatic change on water in the 18 water regions of the conterminous United States defined by the U.S. Water Resources Council (1978). They find that a 2°C warming and a 10% reduction in precipitation would not have serious effects in the humid regions east of the 100th meridian. In the West, however, the impact would be severe on seven water regions: the drainage basins of the Missouri, Arkansas–White–Red, Rio Grande, and Colorado rivers; the river basins draining into the Gulf of Mexico from the northern two-thirds of Texas; and the rivers of California. The only western water regions that would not be severely affected are the water-rich Pacific Northwest and the Great Basin (parts of Nevada, Utah, and Idaho), where demand is relatively small and groundwater reserves are large.

In estimating the impact of climate change, Stockton and Boggess assumed:

1. The region-by-region variation in annual runoff is predominantly influenced by climate, although other factors such as geology, topography, vegetation, and many other variables may be important, especially in smaller drainages.
2. The empirical curves associating total annual precipitation and total annual runoff with weighted mean annual temperature are appropriate for all 18 regions although derived from a relatively small (22 drainage basin) sample.
3. Changes in land use have relatively small influences on regionwide annual runoff.
4. Annual runoff is not greatly affected by large-scale groundwater overdraft.

5. Evapotranspiration is controlled solely by temperature.
6. The postulated climatic change does not modify the present monthly distribution of temperature and precipitation; only the amplitude of the present distribution is increased or decreased.
7. Selection of a few meteorological stations for each region adequately establishes the relation of the weighted mean temperature and annual precipitation to annual runoff.

In Figure 11.4 we have summarized the estimates by Stockton and Boggess of the effects of a 2°C increase in temperature and a 10% reduction in precipitation in the seven water regions of the western United States in which climate change would have the most serious impact. These regions cover about half the area of the conterminous United States, but they produce only about 15% of the mean annual stream runoff. The table shows the present mean annual water supply for each region, not including mined groundwater, in millions of hectare meters (10^{10} m^3 yr^{-1}) and the estimated mean annual requirement in the year 2000. The mean annual requirements listed in Figure 11.4 represent "consumptive" use of water plus evaporation from reservoirs, that is, the total quantity of water that is evapotranspired in the course of beneficial human use. Although actual withdrawals from streams and underground aquifers are considerably larger, portions of these withdrawals are returned to the streams or back into the ground where the water may be reused. In the present climate the ratio of estimated requirements to supplies is less than one for all regions except the Lower Colorado River. In this region today, the deficit of supply is presently made up by extensive mining of groundwater.

For the postulated climatic change, supplies would greatly diminish in all regions, ranging from almost a 76% reduction in the Rio Grande region to nearly 40% in the Upper Colorado, with the result that estimated requirements would exceed supplies in the Missouri, Rio Grande, and Upper and Lower Colorado regions. Mean annual requirements would still be less than future mean annual supplies in the Arkansas–White–Red, Texas Gulf, and California regions. But requirements would almost certainly exceed supplies in the Texas Gulf and California regions during future prolonged droughts. Conditions are highly variable in different parts of the Arkansas–White–Red region, with the western part tending to be deficient in water supplies and the eastern part having a surplus. To maintain the present pattern of water

FIGURE 11.4

Comparison of Water Requirements and Supplies for Present Climatic State and for a 2°C Increase in Temperature and 10% Reduction in Precipitation in Seven Western U.S. Water Regions[a]

Water Region[b]	Area ($10^{10}/m^2$)	Mean Annual Runoff ($10^{10}\ m^3\ yr^{-1}$)	(mm)	Present Climate			Warmer and Drier Climate		
				Mean Annual Supply ($10^{10}\ m^3\ yr^{-1}$)	Mean Annual Requirements[c] ($10^{10}\ m^3\ yr^{-1}$)	Ratio of Requirement to Supply	Mean Annual Supply ($10^{10}\ m^3\ yr^{-1}$)	Percent Change in Supply	Ratio of Requirement[d] to Supply
Missouri	132.4	8.50	64	8.50	3.63	0.43	3.07	−63.9	1.18
Arkansas–White–Red	63.2	9.35	148	9.35	1.67	0.18	4.32	−53.8	0.39
Texas Gulf	44.9	4.92	110	4.92	1.74	0.35	2.47	−49.8	0.70
Rio Grande	35.2	0.74	21	0.74	0.67	0.91	0.18	−75.7	3.72
Upper Colorado	29.6	1.64[e]	55	1.64	1.63[f]	0.99	0.99	−39.6	1.65
Lower Colorado	40.1	0.38	10	1.15[g]	1.37	1.19	0.50[g]	−56.5	2.68
California	42.9	9.56	222	10.18[h]	4.22	0.41	5.71[g]	−43.9	0.74
For the 7 regions together	388.3	35.09	90.4	35.09[h]	14.93	0.43	16.53[h]	−53	0.90

[a]SOURCE: Stockton and Bogess (1979) and calculations in this paper for Upper Colorado Basin.

[b]As defined by the U.S. Water Resources Council (1978).

[c]Projected through year 2000 A.D.

[d]Assuming no increase in requirement because of increased evapotranspiration from irrigated farms or reservoirs.

[e]Average "virgin flow" of the Colorado River at Lee Ferry from 1931 to 1976.

[f]Includes allocation to Lower Basin States, California included, of $0.93 \times 10^{10}\ m^3\ yr^{-1}$.

[g]Includes water received from Upper Colorado Basin, but not mined groundwater.

[h]Total is less than sum of the column because of flow of Lower Colorado derived from Upper Colorado (g).

use, large-scale transfers between basins might be necessary here even under average conditions, let alone to meet water requirements during prolonged droughts. A serious deterioration in water quality would follow from climatic change in all seven regions.

The ratio of future requirements to supply would probably be even less favorable than indicated in Figure 11.4 because evapotranspiration from irrigated farms and reservoirs would undoubtedly increase with a rise in temperature. On a global basis this would be compensated for by an increase in precipitation, but this might or might not occur in the regions we are considering.

At present, California depends for about 15% of its water on imports from the Colorado River. These imports might be eliminated entirely with the postulated climatic change, in which case the ratio of mean annual requirements to mean annual runoff would increase to 0.83, more than double the present ratio.

In all seven regions, irrigation is by far the largest user. Its share of water withdrawals ranges from 68% in the Texas Gulf region to 95% in the Rio Grande region. Total water withdrawals for agriculture are now 13.2 million hectare meters, and in the seven regions the irrigated area is (very approximately) 13 million hectares (Rogers, 1983) so that, on the average, the annual depth of irrigation is about 1 m. Consumptive water use in irrigation is much smaller; a large share of the water that is not consumed reappears as return flows that can be used downstream. Reduction in the irrigated area and an increase in the efficiency of water use in irrigation would significantly lower the overall water require-ments in the seven western regions. A 15% increase in water-use efficiency is probably feasible (Jensen, 1982). Reduction in the irrigated area might come about automatically if an interstate economic market for water were to develop, because economic returns to irrigation are relatively low in large parts of the seven western regions. On the other hand, potentially very large Indian claims for irrigation water for their reserved lands must eventually be settled, and this could result in a major reallocation of water rights (Back and Taylor, 1980).

The effects of future droughts in the Arkansas–White–Red, Texas Gulf, and California regions from the assumed climatic change could be significantly mitigated by construction of additional reservoirs for water storage, but increases in storage would help little in the Missouri, Rio Grande, or Upper and Lower Colorado regions because their storage reservoirs are already so large compared to the annual runoff. In these regions strict water conservation would be essential.

The mean annual requirements listed in Figure 11.4 represent "consumptive" use of water plus evaporation from reservoirs, that is, the total quantity of water that is evapotranspired in the course of beneficial human use. Although actual withdrawals from streams and underground aquifers are considerably larger, portions of these withdrawals are returned to the streams or back into the ground where the water may be reused.

As Stockton and Boggess show, the one western region where a large surplus would still exist after their postulated climatic change would be the Pacific Northwest. Their estimated ratio of requirements to supplies following a 2°C warming and a 10% reduction in precipitation would be 0.10. The annual supply would then be 23.7 million hectare meters. Transfer of 20% of this total supply to water-short regions through large, long-distance conveyance could increase future supplies in the seven western regions shown in Figure 11.4 by nearly 30%, thereby compensating for much of the estimated shortages from climatic change. The ratio of requirements to supplies in the Pacific Northwest region would still be a comfortable 0.18.

From an economic standpoint, however, such a transfer would probably not be desirable. The value of the hydroelectric energy that could be generated from a hectare meter of water in the Pacific Northwest, assuming a total head of 500 m and a price of 5¢ per kilowatt-hour, would be over $600, considerably in excess of the value of a hectare meter of water for irrigating the fodder and cereal crops grown in most of the western regions (Rogers, 1983). . . .

CLIMATE CHANGE AND WATER-RESOURCE SYSTEMS

Planning and construction of major water-resource systems have a time constant of 30 to 50 years. In the past, these activities have been based on the explicit assumption of unchanging climate. The probably serious economic and social consequences of a carbon dioxide-induced climatic change within the next 50 to 100 years warrant careful consideration by planners of ways to create more robust and resilient water-resource systems that will, insofar as possible, mitigate these effects.

References

Back, W. D., and J. S. Taylor (1980). Navajo water rights: pulling the plug on the Colorado River. *Natural Resources Journal,* January 1980, pp. 70–90.

Dracup, J. A. (1977). Impact on the Colorado River Basin and Southwest water supply. In *Climate, Climatic Change, and Water Supply.* National Academy of Sciences, National Academy Press, Washington, D.C.

Howe, C. W., and A. H. Murphy (1981). The utilization and impacts of climate information on the development and operations of the Colorado River system. In *Managing Climatic Resources and Risks.* Panel on the Effective Use of Climate Information in Decision Making, Climate Board, National Academy of Sciences, Washington, D.C., pp. 36–44.

Jensen, M. E. (1982). Water resources technology and management. Paper presented at Soil and Water Resources Conservation Act Symposium on Future Agricultural Technology and Resource Conservation, December 1982.

Laney, N. (1982). Does Arizona's 1980 Groundwater Management Act violate the Commerce Clause? *Arizona Law Review 24:*108–131.

Langbein, W. B., et al. (1949). *Annual Runoff in the United States.* U.S. Geological Survey Circular 5. U.S. Dept. of the Interior, Washington, D.C. (Reprinted 1959.)

Manabe, S., and R. T. Wetherald (1980). On the horizontal distribution of climate change resulting from an increase of CO_2 content of the atmosphere. *J. Atmos. Sci. 37:*99–118.

Rogers, P. (1983). Water resource technology and management in the future of U.S. agriculture. Paper presented at the Soil and Water Resources Conservation Act Symposium on Future Agricultural Technology and Resource Conservation, December 1982.

Schaake, Jr., J. S., and Z. Kaczmarek (1979). Climate variability and the design and operation of water resource systems. In *Proceedings of the World Climate Conference,* pp. 290–312. World Meteorological Organization, Geneva, Switzerland.

Stockton, C. W., and W. R. Boggess (1979). Geohydrological implications of climate change on water resource development. U.S. Army Coastal Engineering Research Center, Fort Belvoir, Virginia.

U.S. Water Resources Council (1978). *The Nation's Water Resources: The Second National Water Assessment.* U.S. Government Printing Office, Washington, D.C.

Chapter 12

THE CAUSES AND EFFECTS OF SEA LEVEL RISE

James G. Titus

INTRODUCTION

For the last several thousand years, sea level has risen so slowly that for most practical purposes it has been constant. As a result, people and other maritime species have had the opportunity to extensively develop the shorelines of the world. Whether one is talking about a vacation spot in Rio de Janeiro, swamps in Bangladesh, farmland in the Nile Delta, marshes along the Chesapeake Bay, or the merchants of Venice, life along the coast is in a sensitive balance with the level of the sea.

This balance would be upset by the rise in sea level that could result from the global warming projected by Hansen et al. Such a warming

This chapter originally appeared in *Effects on Changes in Stratospheric Ozone and Global Climate*, Volume 1, published in 1986 by the Environmental Protection Agency. The original publication of this paper provided an illustration that has not been included here. This illustration depicted computed near-equilibrium changes in ocean temperature for a doubling of atmospheric carbon dioxide and probable increase in other greenhouse gases for approximately the year 2080.

FIGURE 12.1
Snow and Ice Components

	Area (10^6 km²)	Ice Volume (10^6 km³)	Sea Level Equivalent[a] (m)
Land ice: East Antarctica[b]	9.86	25.92	64.8
West Antarctica[c]	2.34	3.40	8.5
Greenland	1.7	3.0	7.6
Small ice caps and mountain glaciers			
(Hollin and Barry 1979;	0.54	0.12	0.3
Flint 1971)			0.6
Permafrost (excluding Antarctica):			0.6
Continuous	7.6	0.03	0.08
		to	to
Discontinuous	17.3	0.7	0.17
Sea ice: Arctic[d]			
Late February	14.0	0.05	
Late August	7.0	0.02	
Antarctic[e]			
September	18.4	0.06	
February	3.6	0.01	
Land Snow Cover[f]			
N. Hemisphere			
Early February	46.3	0.002	
Late August	3.7		
S. Hemisphere			
Late July	0.85		
Early May	0.07		

[a]400,000 km³ of ice is equivalent to 1 m global sea level.

[b]Grounded ice sheet, excluding peripheral, floating ice shelves (which do not affect sea level). The shelves have a total area of 1.62×10^6 km² and a volume of 0.79×10^6 km³ (Drewry and Heim 1983).

[c]Including the Antarctic Peninsula.

[d]Excluding the Sea of Okhotsk, the Baltic Sea, and the Gulf of St. Lawrence (Walsh and Johnson 1979). Maximum ice extents in these areas are 0.7 million, 0.4 million, and 0.2 million km², respectively.

[e]Actual ice area excluding open water (Zwally et al. 1983). Ice extent ranges between 4 million and 20 million km².

[f]Snow cover includes that on land ice but excludes snow-covered sea ice (Dewey and Heim 1981).

(*Modified from Hollin and Barry, 1979. Source:* Glaciers, Ice Sheets, and Sea Level, *National Academy Press, p. 272.*)

could raise sea level one meter or more in the next century by expanding ocean water, melting mountain glaciers, and perhaps eventually causing polar ice sheets to melt or slide into the oceans. sea level rise would inundate low-lying areas, drown coastal marshes and swamps, erode beaches, exacerbate flooding, and increase the salinity of rivers, bays, and aquifers throughout the world.

This paper provides an overview of the causes and effects of sea level rise.

CAUSES OF SEA LEVEL RISE

PAST TRENDS IN SEA LEVEL

The worldwide average sea level depends primarily on (a) the shape and size of ocean basins, (b) the amount of water in the oceans, and (c) the average density of seawater. Subsidence, emergence, and other local factors can cause trends in "relative sea level" at particular locations to differ from trends in "global sea level."

Hays and Pitman (1973) analyzed fossil records and concluded that over the last 100 million years, changes in mid-ocean ridge systems have caused sea level to rise and fall over 300 meters. However, Clark, Farrell, and Peltier (1978) have pointed out that these changes have accounted for sea level changes of less than one millimeter per century. No published study has indicated that this determinant of sea level is likely to have a significant impact in the next century.

The impact of climate on sea level has been more pronounced. Geologists generally recognize that during ice ages the glaciation of substantial portions of the Northern Hemisphere has removed enough water from the oceans to lower sea level one hundred meters below present levels during the last (18,000 years ago) and previous ice ages (Don, Farrand, and Ewing, 1962; Kennett, 1982; Oldale, 1985).

Although the glaciers that once covered much of the Northern Hemisphere have retreated, the world's remaining ice cover contains enough water to raise sea level over seventy-five meters (Hollin and Barry, 1979). Figure 12.1 shows estimates by Hollin and Barry (1979) and Flint (1971) that existing alpine glaciers contain enough water to raise sea level 30 or 60 centimeters, respectively. The Greenland and West Antarctic ice sheets each contain enough water to raise sea level about seven meters, while East Antarctica has enough ice to raise sea level over 60 meters.

There is no evidence that either the Greenland or East Antarctic ice sheets have completely disintegrated in the last two million years. However, it is generally recognized that sea level was about seven meters higher than today during the last interglacial period, which was one to two degrees warmer (Moore, 1982; Mercer, 1968). Because the West Antarctic ice sheet is marine-based and thought to be vulnerable to climatic warming, attention has focused on this source for the higher sea level. Mercer (1968) found that lake sediments and other evidence suggested that summer temperatures in Antarctica have been 7° to 10°C higher than today at some point in the last two million years, probably during the last interglacial period 125,000 years ago, and that such temperatures could have caused a disintegration of the West Antarctic ice sheet.

Tidal gauges have been available to measure the change in sea level at particular locations over the last century. Studies combining these measurements to estimate global trends have concluded that sea level has risen 1.0 to 1.5 mm/yr during the last century (Barnett, 1983; Gornitz, Lebedeff, and Hansen, 1982; Fairbridge and Krebs, 1962). Barnett (1983) found the rate of sea level rise for the last fifty years to be between 2.0 and 2.5 mm/yr, while in the previous fifty years there was little change; however, the acceleration of the rate of sea level rise was not statistically significant. Emery and Aubrey (1985, n.d.) have filtered out estimated land surface movements in their analyses of tidal gauge records in northern Europe and western North America, and have found an acceleration in the rate of sea level rise over the last century.[1] Braatz and Aubrey (n.d.) have found that the rate of relative sea level rise on the east coast of North America accelerated after 1934.

Several researchers have attempted to explain the source of current trends in sea level. Barnett (1984) and Gornitz, Lebedeff, and Hansen (1982) estimate that thermal expansion of the upper layers of the oceans resulting from the observed global warming of 0.4°C in the last century could be responsible for a rise of 0.4 to 0.5 millimeter per year. Roemmich and Wunsch (1984) examined temperature and salinity measurements at Bermuda, found that the 4°C isotherm had migrated 100 meters downward, and concluded that the resulting expansion of ocean water could be responsible for some or all of the observed rise in relative sea level. Roemmich (1985) showed that the warming trend 700 meters below the surface was statistically significant. However, Barnett (1983) found no significant trend based on an examination of the upper layers of the ocean. Nevertheless, Braatz and Aubrey (n.d.) note that long-term steric changes in the ocean are not confined to the upper layers of the oceans, which implies that the Barnett analysis does not necessarily contradict the Roemmich and Wunsch conclusion.

Meier (1984) estimates that retreat of alpine glaciers and small ice caps could be responsible for a current contribution to sea level of between 0.2 and 0.72 millimeter per year. The National Academy of Sciences (NAS) Polar Research Board (Meier 1985) concluded that existing information is insufficient to determine whether the impacts of Greenland and Antarctica are positive, negative, or zero. Although the estimated global warming of the last century appears at least partly responsible for the last century's rise in sea level, no study has demonstrated that global warming might be responsible for an accelerated rate of sea level rise.

IMPACT OF FUTURE GLOBAL WARMING ON SEA LEVEL

Concern about a substantial rise in sea level as a result of the projected global warming stemmed originally from Mercer (1968), who suggested that the Ross and Filchner-Ronne ice shelves might disintegrate, causing a deglaciation of the the West Antarctic ice sheet and a resulting six to seven meter rise in sea level, possibly within 40 years.

Subsequent investigations have concluded that such a rapid rise is unlikely. Hughes (1983) estimated that such a disintegration would take at least two hundred years, and Bentley (1983), five hundred. Other researchers have estimated that this process could take considerably longer (Fastook, 1985; Lingle, 1985).

Researchers have turned their attention to the magnitude of sea level rise that might occur in the next century. The best understood factors are the thermal expansion of ocean water and the melting of alpine glaciers. In the National Academy of Sciences report *Changing Climate,* Revelle (1983) used the model of Cess and Goldenberg (1981) to estimate temperature increases at various depths and latitudes resulting from a 4.2°C warming by 2050–2060. While noting that his assumed time constant of 33 years probably resulted in a conservatively low estimate, he estimated that temperature increases would result in an expansion of the upper ocean sufficient to raise sea level thirty centimeters.

Using a model of the oceans developed by Lacis et al. (1981), Hoffman, Wells, and Titus (1985) examined a variety of possible scenarios of future emissions of greenhouse gases and global warming. They estimated that a warming of between 1° and 2.6°C could result in a thermal expansion contribution to sea level of between 12 and 26 cm by 2050. They also estimated that a global warming of 2.3° to 7.0°C by 2100 would result in thermal expansion of 28–83 cm by that year.

Revelle (1983) suggested that while he could not estimate the future contribution of alpine glaciers to sea level rise, a contribution of 12 centimeters through 2080 would be reasonable. Meier (1984) used glacier balance and volume change data for twenty-five glaciers where the available record exceeded fifty years to estimate the relationship between historic temperature increases and the resulting negative mass balances of the glaciers. He estimated that a 28-mm rise had resulted from a warming of 0.5°C, and concluded that a 1.5° to 4.5°C warming would result in a rise of 8–25 cm in the next century. Using these results, the NAS Polar Board concluded that the contribution of glaciers and small ice caps through 2100 is likely to be 10 to 30 cm (Meier et al., 1985). They noted that the gradual depletion of remaining ice cover might reduce the contribution of sea level rise somewhat. However, the contribution might also be greater, given that the historic rise took place over a sixty-year period, while the forecast period is over one hundred years. Using Meier's estimated relationship between global warming and the alpine contribution, Hoffman, Wells, and Titus (1986) estimated alpine contributions through 2100 at 12–38 cm for a global warming of 2.3° to 7.0°C.

The first published estimate of the contribution of Greenland glacier meltwater to sea level was Revelle's (1983) estimate of 12 cm through the year 2080. Using estimates by Ambach (1980, 1982) that the equilibrium line (between snowfall accumulation and melting) rises one hundred meters for each 0.6°C rise in air temperatures, he concluded that the projected 6°C warming of Greenland would be likely to raise the equilibrium line 1,000 meters. He estimated that such a change in the equilibrium line would result in a 12-cm contribution to sea level rise for the next century.

The NAS Polar Board (Meier et al., 1985) noted that the large ablation area makes Greenland a "significant potential contributor of meltwater to the ocean if climatic warming causes an increase in the rate of ablation and an upward shift of the equilibrium line." They found that a 1,000-m rise in the equilibrium line would result in a contribution of 30 cm through 2100. However, because Ambach (1985) found the relationship between the equilibrium line and temperature to be 77 meters per degree (C), the panel concluded that a 500-m shift in the equilibrium line would be more likely. Based on the assumption of a 6.5°C warming by 2050 and constant temperatures thereafter, the panel estimated that such a change would contribute about 10 cm to sea level through 2100, but also noted that "for an extreme but highly unlikely case, with the equilibrium line raised 1,000 m, the total rise would be 26 centimeters." Although Bindschadler (1985) had treated the two cases as equally plausible, his analysis was conducted before the results of Ambach (1985) were known; he has since indicated agreement with the findings of Meier et al. (1985).[2]

Available estimates of the Greenland contribution assume that all meltwater flows into the oceans and that the ice dynamics of the glaciers do not change. The NAS Polar Board suggested that some of the water would refreeze, decreasing the contribution to sea level rise. Although a change in ice dynamics might imply additional deglaciation and increase the rate of sea level rise, the panel assumed that such changes were unlikely to occur in the next century.

The potential impact of a global warming on Antarctica in the next century is the least certain of all the factors by which a global warming might contribute to sea level rise. Meltwater from East Antarctica might make a significant contribution by the year 2100, but no one has estimated it.[3] The contribution of ice sliding into the oceans, known as "deglaciation," has been the subject of several studies.

Bentley (1983) examined the processes by which a deglaciation of Antarctica might occur. First an accelerated melting of the undersides of the Ross and Filchner-Ronne ice shelves would occur due to warmer water circulating underneath them. The thinning of these ice shelves could cause them to become unpinned and their grounding lines to retreat. Revelle (1983) suggests that the ice shelves might disappear in 100 years, after which time the Antarctic ice streams would flow directly into the oceans, without the back pressure of the ice shelves.

Bentley suggests that all the ice could be discharged over a period of 500 years.

Although a West Antarctic deglaciation would occur over a period of centuries, it is possible that an irreversible deglaciation could commence before 2050. If the ice shelves thinned more than about one meter per year, Thomas, Sanderson, and Rose (1979) suggested that the ice would move into the sea at a sufficient speed that even a cooling back to the temperatures of today would not be sufficient to result in a reformation of the ice shelf.

To estimate the likely Antarctic contribution for the next century, Thomas (1985) developed four scenarios measuring the impact of a 3°C global warming by 2050:

- A shelf melting rate of 1 m/yr with seaward ice fronts remaining at present locations—implies a rise of 28 cm by the year 2100.
- A shelf melting rate of 1 m/yr with ice fronts calving back to a line linking the areas where the shelf is grounded, during the 2050s— implies a rise of 1.6 m by 2100.
- Same as case 1 but with a melt rate of 3 m/yr—implies a rise of 1 m by 2100.
- Same as case 2 but with a melt rate of 3 m/yr—implies a rise of 2.2 m by 2100.

Thomas concluded that the 28-cm rise implied by case 1 would be most likely. He also stated that even if enhanced calving did occur, it would be likely to occur after 2050, "suggesting that probably associated sea level rise would be closer to the 1 m of case 3 than the 2.2 m of case 4."

The NAS Polar Board (Meier et al., 1985) evaluated the Thomas study and papers by Lingle (1985) and Fastook (1985). Although Lingle estimated that the contribution of West Antarctica through 2100 would be 3 to 5 cm, he did not evaluate the contribution from East Antarctica, while Fastook made no estimate for the year 2100. Thus, the panel concluded that "imposing reasonable limits" on Thomas' model yields a range of 20 to 80 cm by 2100 for the Antarctic contribution. However, they also noted several factors that could reduce the amount of ice discharged into the sea: the removal of the warmest ice from the ice shelves, the retreat of grounding lines, and increased lateral shear stress. They also concluded that increased precipitation over Antarctica *might* increase the size of the polar ice sheets there. Thus, the panel

FIGURE 12.2
Estimates of Future Sea Level Rise (centimeters)

Year 2100 by Cause (2085 in the case of NAS 1983)

	Thermal Expansion	Alpine Glaciers	Greenland	Antarctica	Total
NAS (1983)	30	12	12*	70	
EPA (1983)	28–115	#	##	56–345	
NAS (1985)##	—	10–30	10–30	− 10 − + 100	50–200
Thomas (1985)	—	—	—	0–200	—
Hoffman et al. (1986)	28–83	12–37	6–27	12–220	57–368

*Total Rise in Specific Years:***

	2000	2025	2050	2075	2085	2100
NAS (1983)	—	—	—	−70	—	
EPA (1983)						
low	4.8	13	23	38	—	56.0
mid-range low	8.8	26	53	91	—	144.4
mid-range high	13.2	39	79	137	—	216.6
high	17.1	55	117	212	—	345.0
Hoffman et al. (1986)						
low	3.5	10	20	36	44	57
high	5.5	21	55	191	258	368

*Revelle (1983) attributes 16 cm to other factors.
**Only EPA reports made year-by-year projections for the next century.
#Hoffman et al. (1983) assumed that the glacial contribution would be one to two times the contribution of thermal expansion.
##NAS (1985) estimate includes extrapolation of thermal expansion from Revelle (1983).

(*Sources: Hoffman et al., 1986; Meier et al., 1985; Hoffman et al., 1983; Revelle, 1983; Thomas, 1985.*)

concluded that Antarctica could cause a rise in sea level up to 1 m, or a drop of 10 cm, with a rise between 0 and 30 cm most likely.

Figure 12.2 summarizes the various estimates of global sea level rise. The report *Projecting Sea Level Rise* (Hoffman, Keyes, and Titus, 1983) estimated that the rise would be between 56 and 345 cm, with a probable rise between 144 and 217 cm. Revelle (1983) estimated that the rise was likely to be 70 cm, ignoring the impact of a global warming on Antarctica; Revelle also noted that the latter contribution was likely to be 1 to 2 m per century after 2050, but declined to add that to his estimate. The NAS Polar Board (Meier et al., 1985) projected that the contribution of glaciers would be sufficient to raise sea level 20 to 160 cm, with a rise of "several tenths of a meter" most likely. Thus, if one extrapolates the earlier NAS estimate of thermal expansion through the year 2100, the 1985 NAS report implies a rise between 50 and 200 cm. The estimates of Hoffman, Wells, and Titus (1985) were similar to the estimates of Hoffman, Keyes, and Titus (1983) for the year 2100, but for the year 2025, they lowered their estimate from 26–39 cm to 10–21 cm.

FUTURE TRENDS IN LOCAL SEA LEVEL

Although most attention has focused on projections of global sea level, impacts on particular areas would depend on local relative sea level. Tidal gauge measurements suggest that relative sea level has risen 10 to 20 cm per century more rapidly than the worldwide average along much of the U.S. coast (Hicks, DeBaugh, and Hickman, 1983). However, Louisiania is subsiding close to 1 m per century, while parts of Alaska are emerging 10 or more cm per century. Bruun argues that throughout most of the world, sea level has been rising. However, Bird (Titus, 1986) argues that sea level appears to be stable in Australia, which may imply that future rates of sea level rise will also be 10 to 15 cm per century less than the worldwide average.

Local subsidence and emergence are caused by a variety of factors. Rebound from the retreat of glaciers after the last ice age has resulted in the emergence of Alaska and parts of Scandinavia. The emergence in polar latitudes has resulted in subsidence in other areas. Groundwater pumping has caused rapid subsidence around Houston, Texas; Taipai, Taiwan; and Bangkok, Thailand, among other areas (Leatherman, 1984; Kuo, Titus, 1986). River deltas and other newly created land subside as the unconsolidated materials compact.

Although subsidence and emergence trends may change in the future, particularly where anthropogenic causes are curtailed, no one has linked these causes to future climate change in the next century. However, the removal of ice from Greenland and Antarctica would deform the ocean floor. Clark and Lingle (1977) have calculated the impact of a uniform 1-m contribution from West Antarctica. They concluded that relative sea level at Hawaii would increase by an additional 25 cm, and that along much of the U.S. Atlantic and Gulf Coasts there would be an additional 15 cm. On the other hand, sea level would drop at Cape Horn by nearly 10 cm, and the rise along the southern half of the Argentine and Chilean coasts would be less than 75 cm.

Other influences on local sea level that might change as a result of a global warming include currents, winds, and freshwater flow into estuaries. None of these impacts, however, has been estimated.

EFFECTS OF SEA LEVEL RISE

A rise in sea level of 1 or 2 m would permanently inundate wetlands and lowlands, accelerate coastal erosion, exacerbate coastal flooding, threaten coastal structures, and increase the salinity of estuaries and aquifers. Substantial research has been done on the implications of sea level rise for coastal erosion and wetlands, while relatively little work has been done in the other areas.

SUBMERGENCE OF COASTAL WETLANDS

The most direct impact of a rise in sea level is the inundation of areas that had been just above the water level before the sea rose. Coastal wetlands are generally found at elevations below the highest tide of the year and above mean sea level. Thus, wetlands account for most of the land less than 1 m above sea level.

Because a common means of estimating past sea level rise has been the analysis of marsh peats, the impacts of sea level rise on wetlands are fairly well understood. For the rates of sea level rise of the last several thousand years, marshes have generally kept pace with sea level through sedimentation and peat formation (Emery and Uchupi, 1972; Redfield, 1972, 1967; Davis, 1985). As sea level rose, new wetlands formed inland while the seaward boundary was maintained (Figure 12.3a and 12.3b). Because the wetland area has expanded, Titus, Henderson, and Teal (1984) hypothesized that one would expect a concave marsh profile, i.e., that there is more marsh area than the area found immediately above the marsh. Thus, if sea level rose more rapidly than the marsh's ability to keep pace, there would be a net loss of wetlands (Figure 12.3c). Moreover, a complete loss might occur if protection of developed areas prevented the inland formation of new wetlands (Figure 12.3d).

FIGURE 12.3
Evolution of Marsh as Sea Level Rises

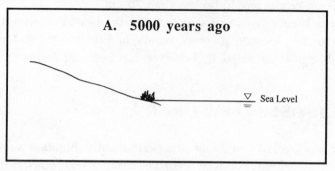

A. 5000 years ago

Sea Level

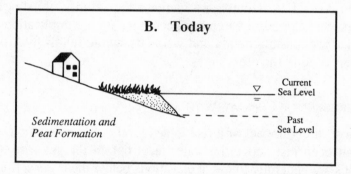

B. Today

Current
Sea Level

Past
Sea Level

*Sedimentation and
Peat Formation*

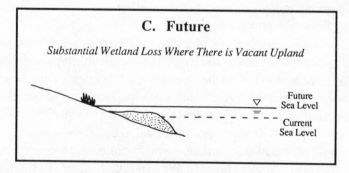

C. Future

Substantial Wetland Loss Where There is Vacant Upland

Future
Sea Level

Current
Sea Level

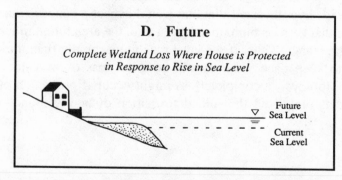

D. Future

*Complete Wetland Loss Where House is Protected
in Response to Rise in Sea Level*

Future
Sea Level

Current
Sea Level

Kana, Baca, and Williams (1986) and Kana et al. (n.d.) surveyed marsh transects in the areas of Charleston, South Carolina, and two sites near Long Beach Island, New Jersey, to evaluate the concavity of wetland profiles and the vulnerability of wetlands to a rise in sea level. Their data from the Charleston area showed all the marsh to be between 30 and 110 cm above current sea level, an elevation range of 80 cm. The area with a similar elevation range just above the marsh was only 20% as large. Thus, a rise in sea level exceeding vertical marsh accretion by 80 cm would result in an 80% loss of wetlands. In the New Jersey sites, the marsh was also found within an elevation range of 80 cm; a rise in sea level 80 cm in excess of marsh accretion would result in 67 to 90% losses.

The future ability of marshes to accrete vertically is uncertain. Based on field studies by Ward and Domeracki (1978), Hatton et al. (1983), Meyerson (1972), and Stearns and MacCreary (1957), Kana et al. (n.d.) concluded that current vertical accretion rates are approximately 4 to 6 mm/yr in the two case study areas, greater than the current rate of sea level rise but less than the rates of rise projected for the next century. If current accretion trends continue, then 87 and 160 cm rises by 2075 would imply 50 and 80% losses of wetlands in the Charleston area. Kana et al. (n.d.) also estimated 80% losses in the New Jersey sites for a 160-cm rise through 2075. However, because the high marsh dominates in that area, they concluded that the principal impact of an 87-cm rise by 2075 would be the conversion of high marsh to low marsh.

In both cases, the losses of marsh could be greater if inland areas are developed and protected with bulkheads or levees. Because there is a buffer zone between developed areas and the marsh in South Carolina, protecting development from a 160-cm rise would increase the loss from 80–90%. Without the buffer, the loss would be close to 100%.

Louisiana, whose marshes and swamps account for 40 percent of the coastal wetlands in the United States (excluding Alaska), would be particularly vulnerable to an accelerated rise in sea level. The majority of the Louisiana wetlands are less than one meter above sea level, and are generally subsiding approximately one meter per century as its deltaic sediments compact (Boesch, 1982). Until the last century, the wetlands kept pace with this rate of relative sea level rise, because of the sediment the Mississippi River conveyed to the wetlands.

Human activities, however, have largely disabled the natural processes by which coastal Louisiana might keep pace with sea level rise. Dams, navigation channels, canals, and flood protection levees have interrupted the flow of sediment, fresh water, and nutrients to the wetlands. As a result, over 100 square kilometers of wetlands convert to open water every year (Gagliano et al., 1983). A substantial rise in sea level would further accelerate the process of wetland loss in Louisiana.

To develop an understanding of the potential nationwide impact of sea level rise on coastal wetlands in the United States, Park, Armentano, and Cloonan (Titus, 1986) use topographic maps to characterize wetland elevations at fifty-two sites comprising 4800 square kilometers (1.2 million acres) of wetlands, over 17 percent of all U.S. coastal wetlands (see also Armentano, Park, and Cloonan, n.d.). Using published vertical accretion rates, they estimate the impact of 1.4 and 2.1 m rises in sea level through the year 2100 for each of the sites. Park, Armentano, and Cloonan (Titus, 1986) estimate that these scenarios imply losses of 40 and 76 percent of the existing coastal wetlands in their sample, which could be reduced to 22 and 58% if new wetlands are allowed to form inland. However, Titus (n.d.) found that if the Park et al. sample is weighted according to an inventory of wetlands in particular areas (Alexander, Broutman, and Field, 1986), the resulting estimates of U.S. wetland loss are somewhat higher, 47–82% of existing wetlands, with a potential for reducing those losses to 31 to 70%.

Throughout the world, people have dammed, leveed, and channelized major rivers, curtailing the amount of sediment that reaches river deltas. Even at today's rate of sea level rise, substantial amounts of land are converting to open water in Egypt and Mexico (Milliman and Meade, 1983). Other deltas, such as the Ganges in Bangladesh and India, are currently expanding seaward. These areas would require increased sediment, however, to keep pace with an accelerated rise in sea level. Additional projects to divert the natural flow of river water would increase the vulnerability of these areas to a rise in sea level. Broadus et al. (Titus, 1986) examine this issue in detail for Egypt and Bangladesh. . . .

Several options have been identified for reducing wetland loss due to sea level rise. Abandonment of developed areas inland of today's wetlands could permit new wetlands to form inland. In some cases, it might be possible to enhance the ability of wetlands to accrete vertically by spraying sediment on them or—in the case of Louisiana and other deltas—restoring the natural processes that would provide sediment to the wetlands. Finally, some local governments in Louisiana have proposed to artificially control water levels through the use of levees and pumping stations (Edmonson and Jones, 1985).

The need for anticipating sea level rise would vary. Artificial means to accelerate wetland accretion need not be implemented until the rise takes place (although a lead time would be necessary to develop the required technologies). Similarly, levees and pumping stations could be delayed. On the other hand, a planned retreat would require several decades of lead time to permit the design of new mobile structures and the depreciation of the old immobile structures.

INUNDATION

Although coastal wetlands are found at the lowest elevations, inundation of lowland could also be important in some areas, particularly if sea level rises at least one meter. Unfortunately, the convention of ten-foot contours in the mapping of most coastal areas has prevented a general assessment of land loss. Although a few case studies have been conducted in the United States, very few studies have been undertaken to quantify the potential impacts on other countries, other than the paper by Broadus et al. (Titus, 1986).

Kana et al. (1984) used data from aerial photographs to assess elevations in the area around Charleston. They concluded that 160- and 230-cm rises would result in 30 and 46% losses of the area's dry land, respectively. Leatherman (1984) estimated that such rises would result in 9 and 12% losses of the land in the area of Galveston and Texas City, Texas, assuming that the elaborate network of seawalls and levees were maintained. (Many of the summary results from Leatherman, 1984; Kana et al., 1984; and Gibbs, 1984, appear in the appendix of Titus et al., 1984.)

Schneider and Chen (1980) conducted the first nationwide assessment of the inundation from projected sea level rise. Unfortunately, the smallest rise in sea level they considered was a 4.5-m (15-ft) rise, in part because smaller contours are not generally available in topographic maps. Nevertheless, their findings suggest which coastal states would be most vulnerable: Louisiana (which would lose 28% of its land and 51% of its wealth), Florida (24 and 52%), Delaware (16 and 18%),

Washington, D.C. (15 and 15%), Maryland (12 and 5%), and New Jersey (10 and 9%).

As with wetland loss, the responses to inundation fall broadly into the categories of retreat and holding back the sea. Levees are used extensively in the Netherlands and New Orleans to prevent the flooding of areas below sea level and could be similarly constructed around other major cities. In sparsely developed areas, however, the cost of a levee might be greater than the value of the property being protected. Moreover, even where levees prove to be cost-effective, the environmental implications of replacing natural shorelines with manmade structures would need to be considered.

COASTAL EROSION

Sea level rise can also result in the loss of land above sea level through erosion. Bruun (1962) showed that the erosion resulting from a rise in sea level would depend upon the average slope of the entire beach profile extending from the dunes out to the point where the water is too deep for waves to have a significant impact on the bottom (generally a depth of about 10 meters). By comparison, inundation depends only on the slope immediately above the original sea level. Because beach profiles are generally flatter than the portion of the beach just above sea level, the "Bruun Rule" generally implies that the erosion from a rise in sea level is several times greater than the amount of land directly inundated. (See Figure 12.4.)

As Bird (Titus, 1986) emphasizes, processes other than sea level rise also contribute to erosion, including storms, structures, currents, and alongshore transport. Because sea level has risen slowly in recent centuries, verification of the Bruun Rule on the open coast has been difficult. However, water levels along the Great Lakes can fluctuate over one meter in a decade. Hands (1976, 1979, and 1980) and Weishar and Wood (1983) have demonstrated that the Bruun Rule generally predicts the erosion resulting from rises in water levels there.

The Bruun Rule has been applied to project erosion due to sea level rise for several areas of the United States where it is believed to adequately project future erosion. Bruun (1962) found that a 1-cm rise in sea level would generally result in a 1-m shoreline retreat, but that the retreat could be as great as 10 m along some parts of the Florida coast.

FIGURE 12.4
The Bruun Rule

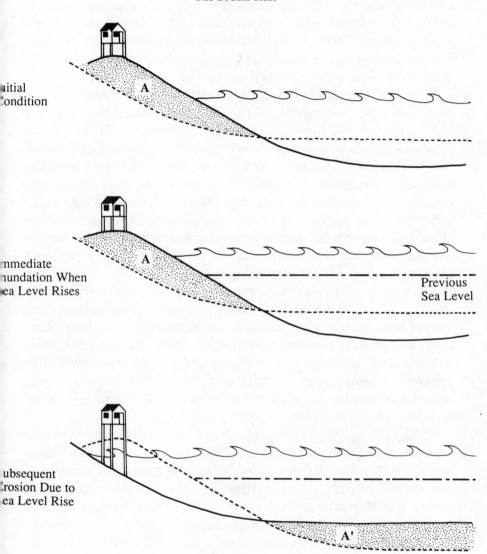

Initial
Condition

Immediate
Inundation When
Sea Level Rises

Previous
Sea Level

Subsequent
Erosion Due to
Sea Level Rise

Everts (1985) and Kyper and Sorensen (1985), however, found that along the coasts of Ocean City, Maryland, and Sandy Hook, New Jersey, respectively, the shoreline retreat implied by the Bruun Rule would be only about 75 cm. Kana et al. (1984) found that along the coast of South Carolina, the retreat could be 2 m. The U.S. Army Corps of Engineers (1979) indicated that along the coast of San Francisco, where waves are generally larger than along the Atlantic Coast, the shore might retreat 2–4 m for a 1-cm rise in sea level.

Dean and Maurmeyer (1983) generalized the "Bruun Rule" approach to consider the "overwash" of barrier islands. Geologists basically believe that coastal barriers can maintain themselves in the face of slowly rising sea level through the landward transport of sand, which washes over the island during storms, building the island upward and landward. Because this formulation of the Bruun Rule extends the beach profile horizontally to include the entire island as well as the active surf zone, it always predicts greater erosion than the Bruun Rule. However, the formulation may not be applicable to developed barrier islands, where the common practice of public officials is to bulldoze sand back onto the beach after a major storm.

The potential erosion from a rise in sea level could be particularly important to recreational beach resorts, which include some of the world's most economically valuable and intensively used land. Relatively few of the most densely developed resorts have beaches wider than about 30 m at high tide. Thus, the rise in relative sea level of 30 centimeters projected in the next 40 to 50 years could erode most recreational beaches in developed areas, unless additional erosion response measures are taken.

Bruun (Titus, 1986) examines potential responses to erosion in considerable detail (see also Magness, 1984; U.S. Army Corps of Engineers, 1977). The responses fall generally into three categories: construction of walls and other structures, the addition of sand to the beach, and abandonment. Although seawalls have been used in the past, they are becoming increasingly unpopular among shore communities because erosion can proceed up to the wall, resulting in a complete loss of beach, which has happened in many areas (Kyper and Sorensen, 1985; Howard, Pilkey, and Kaufman, 1985). A number of other structures have been used to decrease the ability of waves to cause erosion, including groins (jetties) and breakwaters. Bulkheads are often used where waves are small (Sorensen, Weisman, and Lennon, 1984).

A more popular form of erosion control has been the placement of sand onto the beach. Although costs can exceed one million dollars per kilometer (U.S. Army Corps of Engineers, 1980; Howard, Pilkey, and Kaufman, 1985), it is often justified by the economic and recreational value of beaches. A recent study of Ocean City, Maryland, for example, concluded that the cost of holding back the sea for a 30-cm rise in sea level would be about 25 cents per visitor, less than 1 percent of the cost of a trip to the beach (Titus, 1985). That community also provides an example of the practical consequences of sea level rise. Until 1985, the State of Maryland's policy for erosion control was the construction of groins, which curtail erosion caused by sand moving along the shore, but not erosion caused by sea level rise. Sea level rise was cited as the motivating concern for the state to abandon the groin plan and use beach replenishment, which can effectively control erosion caused by both types of erosion (Associated Press, 1985).

Although shore protection is often cost-effective today, the favorable economics might change in the future. A more rapid rise in sea level would increase the costs of shore protection. A number of states have adopted erosion policies that assume a retreat from the shore. North Carolina requires homes that can be moved to be set back from the shore by a distance equal to shoreline recession from 30 years of erosion, while high-rises must be set back 60 years.[4] Maine requires people to demonstrate that new structures will not erode for 100 years.[5] Other jurisdictions discourage the construction of bulkheads and seawalls (Howard, Pilkey, and Kaufman, 1985). As Bird (Titus, 1986) discusses, in many undeveloped countries, small, relatively inexpensive houses are found very close to the shore. Because the value of these houses is less than the cost of protecting them, they must be moved as the shore erodes. An accelerated rise in sea level would speed this process of shoreline retreat.

The need for anticipating erosion caused by sea level rise varies. Where communities are likely to adapt to erosion, anticipation can be important. The cost and feasibility of moving a house back depends on design decisions made when the house is built. The willingness of people to abandon properties depends in part on whether they bought land on the assumption that it would eventually erode away or had assumed that the government would protect it indefinitely. Less anticipation is necessary if the shore will be protected; sand can be added to the beach as necessary. Nevertheless, some advanced planning may be necessary for communities to know whether retreat or defending the shore would be most cost-effective.

FLOODING AND STORM DAMAGE

A rise in sea level could increase flooding and storm damages in coastal areas for three reasons: erosion caused by sea level rise would increase the vulnerability of communities; higher water levels would provide storm surges with a higher base to build upon; and higher water levels would decrease natural and artificial drainage.

The impact of erosion on vulnerability to storms is generally a major consideration in projects proposed to control erosion, most of which have historically been funded through the Army Corps of Engineers in the United States. The impact of sea level rise, however, has not generally been considered separately from other causes of erosion.

The impact of higher base water levels on flooding has been investigated for the areas around Charleston, South Carolina, and Galveston, Texas (Barth and Titus, 1984). Kana et al. (1984) found that around Charleston, the area within the 10-year floodplain would increase from 33% in 1980, to 48, 62, and 74% for rises in sea level of 88, 160, and 230 cm, respectively, and that the area within the 100-year floodplain would increase from 63% to 76, 84, and 90% for the three scenarios. Gibbs (1984) estimated that even an 88-cm rise would double the average annual flood damages in the Charleston area (but that flood losses would not increase substantially for higher rises in sea level because shoreline retreat would result in a large part of the community being completely abandoned).

Leatherman (1984) conducted a similar analysis of Galveston Island, Texas. He estimated that the area within the 100-year floodplain would increase from 58% to 94% for an 88-cm rise in sea level, and that for a rise greater than one meter, the Galveston seawall would be overtopped during a 100-year storm. Gibbs estimated that the damage from a 100-year storm would be tripled for a rise of 88 cm.

A wide variety of shore protection measures would be available for communities to protect themselves from increased storm surge and wave damage due to sea level rise (Sorensen, Weisman, and Lennon, 1984). Many of the measures used to address erosion and inundation, including seawalls, breakwaters, levees, and beach restoration, also provide protection against storms. In the case of Galveston, which is already protected on the ocean side by the seawall, Gibbs hypothesized that it might be necessary to completely encircle the developed areas with a levee to prevent flooding from the bay side; upgrading the existing seawall might also be necessary.

Kyper and Sorensen (1985 and n.d.) examined the implications of sea level rise for the design of coastal protection works at Sea Bright, New Jersey, a coastal community that currently is protected by a seawall and has no beach. Because the seawall is vulnerable to even a 10-year storm, the U.S. Army Corps of Engineers and the State of New Jersey have been considering a possible upgrade. Kyper and Sorensen estimated that the cost of upgrading the seawall for current conditions would be $3.5 to $6 million per kilometer of shoreline, noting that if designed properly, the seawall would be useful throughout the next century. However, they estimated that a rise in relative sea level of 30–40 cm would be likely to result in serious damage to the seawall during a major storm, due to higher water levels and the increased wave heights resulting from the erosion of submerged sand in front of the seawall. To upgrade the seawall to withstand a 1-m rise in relative sea level would cost $5.7 to $9 million per kilometer (50% more). They concluded that policymakers would have to weigh the tradeoff between the cost of designing the wall to withstand projected sea level rise and the cost of subsequent repairs and a second overhaul.

In addition to community-wide engineering approaches, measures can also be taken by individual property owners to prevent increased flooding. In 1968, the U.S. Congress created the National Flood Insurance Program to encourage communities to avoid risky construction in flood-prone areas. In return for requiring new construction to be elevated above expected flood levels, the federal government provides flood insurance, which is not available from the private sector. If sea level rises, flood risks will increase. In response, local ordinances will automatically require new construction to be further elevated, and insurance rates on existing properties will rise unless those properties are further elevated. As currently organized, the National Flood Insurance Program would react to sea level rise as it occurred. Various measures to enable the program to anticipate sea level rise have been proposed, including warning policyholders that rates may increase in the future if sea level rises; denying coverage to new construction in areas that are expected to be lost to erosion within the next 30 years; and setting premiums according to the average risk expected over the lifetime of the mortgage (Howard, Pilkey, and Kaufman, 1985; Titus, 1984).

Kuo (Titus, 1986) describes case studies in Charleston, South Carolina, and Fort Walton Beach, Florida, which examined the implications of sea level rise for rainwater flooding and the design of coastal drainage systems. Waddell and Blaylock (n.d.) estimated that a 25-year rainstorm (with no storm surge) would result in no damages for the Gap Creek watershed in Fort Walton Beach. However, a rise in sea level of 30–45 cm would result in damages of $1.1 to $1.3 million in this community of 4,000 residents during a 25-year storm. An upgrade costing $550,000, however, would prevent such damages.

LaRoche and Webb (n.d.), who had previously developed the master drainage plan for Charleston, South Carolina, evaluated the implications of sea level rise for the Grove Street watershed in that community. They estimated that the costs of upgrading the system for current conditions would be $4.8 million, while the cost of upgrading the system for a 30-cm rise would be $5.1 million. If the system is designed for current conditions and sea level rises, the system would be deficient and the city would face retrofit costs of $2.4 million. Thus, for the additional $300,000 necessary to upgrade for a 30-cm rise, the city could ensure that it would not have to spend an additional $2.4 million later. Noting that the decision whether to design now for a rise in sea level depends on the probability that sea level will rise, they concluded that a 3% real social discount rate would imply that designing for sea level rise is worthwhile if the probability of a 30-cm rise by 2025 is greater than 30%. At a discount rate of 10%, they concluded, designing for future conditions is not worthwhile.

INCREASED SALINITY IN ESTUARIES AND AQUIFERS

Although most researchers and the general public have focused on the increased flooding and shoreline retreat associated with a rise in sea level, the inland penetration of salt water could be important in some areas.

As de Sylva (Titus, 1986) describes, a rise in sea level increases the salinity of an estuary by altering the balance between freshwater and saltwater forces. The salinity of an estuary represents the outcome of (1) the tendency for the ocean salt water to completely mix with the estuarine water and (2) the tendency of fresh water flowing into the estuary to dilute the saline water and push it back toward the ocean. During droughts, the salt water penetrates upstream, while during the rainy season, low salinity levels prevail. A rise in sea level has an impact similar to decreasing the freshwater inflow. By widening and deepening the estuary, sea level rise increases the ability of salt water to penetrate upstream.

The implications of sea level rise for increased salinity have only been examined in detail for Louisiana and the Delaware estuary. In Louisiana and other river deltas, saltwater intrusion is causing the conversion of cypress swamps (which cannot tolerate salt water) to openwater lakes, and increasing the salinity levels of fresh and intermediate marshes. Accelerated sea level rise would speed up this process.

The impact of current sea level trends on salinity has been considered in the long-range plan of the Delaware River Basin Commission since 1981 (DRBC, 1981). The drought of the 1960s resulted in salinity levels that almost contaminated the water supply of Philadelphia and surrounding areas. Hull and Tortoriello (1979) found that the 13-cm rise projected between 1965 and 2000 would result in the "salt front" migrating 2 to 4 kilometers farther upstream during a similar drought. They found that a moderately sized reservoir (57 million cubic meters) to augment river flows would be needed to offset the resulting salinity increases.

Hull, Thatcher, and Tortoriello (1986) examined the potential impacts of an accelerated rise in sea level due to the greenhouse warming. They estimated that 73-cm and 250-cm rises would result in the salt front migrating an additional 15 and 40 kilometers, respectively, during a repeat of the 1960s drought. They also found that the health-based 50-ppm sodium standard (equivalent to 73 ppm chloride) adopted by New Jersey would be exceeded 15 and 50% of the time, respectively, and that the EPA drinking water 250-ppm chloride standard would be exceeded over 35% of the time in the latter case.

Lennon, Wisniewski, and Yoshioka (1986) examined the implications of increased estuarine salinity for the Potomac–Raritan–Magothy aquifer system, which is recharged by the (currently fresh) Delaware River and serves the New Jersey suburbs of Philadelphia. During the 1960s drought, river water with chloride concentrations as high as 150 ppm recharged these aquifers. Lennon et al. estimated that a repeat of the 1960s drought with a 73-cm rise in sea level would result in river water with concentrations as high as 350 ppm recharging the aquifer, and that during the worst month of the drought, over one-half of the water recharging the aquifer system would have concentrations greater than 250 ppm. With a 250-cm rise, 98% of the recharge during the worst month of the drought would exceed 250 ppm, and 75% of the recharge would be greater than 1000 ppm.

Hull and Titus (1986) examined the options by which various agencies might respond to increased salinity in the Delaware estuary. They concluded that planned but unscheduled reservoirs would be more than enough to offset the salinity increased from a one-foot (30-cm) rise in sea level, although those reservoirs had originally been intended to meet increased consumption. They noted that construction of the reservoirs would not be necessary until the rise became more imminent. However, they also suggested that, given the uncertainties, it might be advisable today to identify additional reservoir sites, to ensure that future generations retained the option of building additional reservoirs, if necessary.

A rise in sea level could increase salinities in other areas, although the importance of those impacts has not been investigated. Kana et al. (1984) and Leatherman (1984) made preliminary inquiries into the potential impacts on coastal aquifers around Charleston and Galveston, respectively. However, they concluded that in-depth assessments were not worthwhile because the aquifers around Charleston are already salt-contaminated because of overpumping, and pumping of groundwater has been prohibited in the Galveston area as it causes land subsidence. The potential impacts on Florida's Everglades and the shallow aquifers around Miami might be significant, but these have not been investigated.

ECONOMIC SIGNIFICANCE OF SEA LEVEL RISE

Only two studies have estimated a dollar value of the likely impacts of sea level rise for particular nations. Broadhus et al. (Titus, 1986) examine the impacts on Egypt and Bangladesh. Schneider and Chen

(1980) estimated the economic impact on the United States of what was once (but is no longer) thought to be a plausible scenario: rises of 4.6 to 7.6 meters (15 to 25 feet) occurring with little or no warning during the early part of the twenty-first century. They estimated that these scenarios would result in real property losses of $100 to $150 billion, representing 6.2 to 8.4% of all real property in the nation.

The only comprehensive attempt to place a dollar value on the impacts of sea level rise for particular communities was the study by Gibbs (1984) of the Charleston and Galveston areas, summarized in Volume 4. Gibbs' analysis, which considers scenarios ranging from 0.9 to 2.4-meter rises through 2075, estimates what the economic impact would be if actions are taken in anticipation of sea level rise versus the cost of responding to sea level rise as it occurs. Gibbs also modeled how investment decisions might respond to floods and erosion, and explicitly considered community-wide strategies to limit losses, including shore protection and abandonment.

In the Charleston study, Gibbs assumed that in anticipation of sea-level rise, efforts would be made to avoid developing some vacant suburban areas likely to be flooded in the future; that a partial abandonment would take place; and that the existing seawalls protecting Charleston would be elevated to provide additional protection. For a rise of 28–64 cm through 2025, Gibbs estimated that the present value of the cumulative impact would be $280–$1065 million (5–19% of economic activity in the area for the period), which could be reduced to $160–$420 million if sea level rise was anticipated. Most of this impact would result from a 10–100% increase in expected storm damages, although Gibbs also estimated $7–$35 million in losses as a result of erosion. For the period 1980–2075, Gibbs estimated that the economic impacts would be $1250–$2510 million (17–35%) and could be reduced to $440–$1100 million through anticipatory measures. Gibbs performed a similar analysis of the Galveston area, concluding that the impacts of sea level rise through 2025 would represent $115–$360 million (1.1–3.6%) if not anticipated, and $80–$140 million if anticipated.

Other studies can be used to understand the economic significance of particular classes of impacts. As discussed above, a 30-centimeter (1 foot) rise in sea level would erode most recreational beaches back to the first row of houses. The studies cited in our section on erosion indicate that the typical beach profile extends out about 1000 meters, implying that 300,000 cubic meters of sand per kilometer of shoreline are required to raise the beach profile 30 cm. If sand costs are typically $3–$10 per cubic meter, the beach rebuilding costs of a 30-cm rise in sea level would be $1–$3 million per kilometer. If the United States has a few thousand kilometers of recreational beaches, then it would cost billions and perhaps tens of billions of dollars to rebuild these beaches in response to a 30-cm rise in sea level. This estimate considers only the beaches themselves; raising people's lots to avoid inundation would further increase the costs. The U.S. Army Corps of Engineers (1971) estimated that in 1971, 25,000 kilometers of shoreline (exclusive of Alaska, the Great Lakes, and Hawaii) were eroding, of which 17% were "critically eroding," and would require engineering solutions. If 17% of all shorelines require erosion control, that would imply protection of close to 10,000 kilometers of shoreline. Sorensen (1986) describes dozens of engineering options for preventing erosion, the least expensive of which costs $300,000 per kilometer, implying a cost of at least $3 billion for protecting shorelines.

OTHER IMPACTS OF THE GREENHOUSE WARMING

The impacts of sea level rise on coastal areas, as well as their importance, are likely to depend in part on other impacts of the greenhouse warming. Although future sea level is uncertain, there is a general consensus that a global warming would cause sea level to rise; by contrast, the direction of most other changes is unknown.

One of the more certain impacts is that most areas will be warmer. For coastal resorts in mid-latitudes, the beach season would be extended by a number of weeks. For densely developed communities like Ocean City with a three-month peak season, such an extension might increase revenues 10 to 25%, far more than the estimated cost of controlling erosion. Some areas where the ocean is too cold to swim today might have more tolerable water temperatures in the future. Warmer temperatures in general might encourage more people to visit beaches in the summer.

Warmer temperatures might change the ability of wetlands to keep pace with sea level rise. Mangrove swamps, which are the tropical equivalent of salt marshes, generally accrete differently than salt marshes. If warmer temperatures enable mangroves to grow at higher latitudes, the loss of wetlands to sea level rise may be altered. On the other hand, marsh peat formation is generally greater in cooler climates; warmer temperatures might reduce the rate of vertical accretion for these wetlands.

De Sylva (Titus, 1986) suggests that changing climate could alter the frequency and tracks of storms. Because hurricane formation requires water temperatures of 27°C or higher (Wendland, 1977), a global warming might result in an extension of the hurricane season and in hurricanes forming at higher latitudes. Besides increasing the amount of storm damage, increased frequency of severe storms would tend to flatten the typical beach profile, causing substantial shoreline retreat unless additional sand was placed on the beach. A decreased frequency of severe winter storms might have the opposite impact at higher latitudes.

Because warmer temperatures would intensify the hydrologic cycle, it is generally recognized that a global warming would result in increased rainfall worldwide. Thus, rainwater flooding might be increased because of both decreased drainage and increased precipitation. The impact of sea level rise on saltwater intrusion could be offset by decreased drought frequency or exacerbated by increased drought frequency (Rind and Lebedeff, 1984).

CONCLUSIONS

The studies reviewed in this paper appear to support the following conclusions regarding the causes and effects of sea level rise:

CAUSES

- The projected global warming would accelerate the current rate of sea level rise by expanding ocean water, melting alpine glaciers, and, eventually, causing polar ice sheets to melt or slide into the oceans.
- Global average sea level has risen 10 to 15 cm over the last century. Ocean and glacial studies suggest that the rise is consistent with what models would project, given the 0.4°C warming of the past century. However, no cause and effect relationship has been conclusively demonstrated.

- Projected global warming could cause global average sea level to rise 10 to 20 cm by 2025 and 50 to 200 cm by 2100. Thermal expansion could cause a rise of 25 to 80 cm by 2100; Greenland and alpine glaciers could each contribute 10 to 30 cm through 2100. The contribution of Antarctic deglaciation is likely to be between 0 and 100 cm; however, the possibilities cannot be ruled out that (a) increased snowfall could increase the size of the Antarctic ice sheet, thereby offsetting part of the sea level rise from other sources; or (b) meltwater and enhanced calving of the ice sheet could increase the contribution of Antarctica to as much as 2 meters.
- Disintegration of the West Antarctic ice sheet might raise sea level an additional 6 meters over the next few centuries. Glaciologists generally believe that such a disintegration would take at least 300 years, and probably as long as 500 years. However, a global warming might result in sufficient thinning of the Ross and Filcher-Ronne ice shelves in the next century to make the process irreversible.
- Local trends in subsidence and emergence must be added or subtracted to estimate the rise at particular locations.

EFFECTS

- A substantial rise in sea level would permanently inundate wetlands and lowlands, accelerate coastal erosion, exacerbate coastal flooding, and increase the salinity of estuaries and aquifers.
- Bangladesh and Egypt appear to be among the nations most vulnerable to the rise in sea level projected for the next century. Up to 20% of the land in Bangladesh could be flooded with a 2-meter rise in sea level. Although less than 1% of Egypt's land would be threatened, over 20% of the Nile Delta, which contains most of the nation's people, would be threatened.
- A large fraction of the world's coastal wetlands may be lost, threatening some fisheries. A rise in sea level of 1 to 2 meters by 2100 could destroy 50–80% of U.S. coastal wetlands. Although no study has been taken to estimate the worldwide impact, this result is probably representative.
- Erosion caused by sea level rise could threaten recreational beaches throughout the world. Case studies have concluded that a 30-centimeter rise in sea level would result in beaches eroding 20 to 60 meters or more. Because the first row of houses or hotels is often generally less than 20 meters from the shore at high tide, if available studies are representative, then recreational beaches throughout the world would

be threatened by a 30-centimeter rise unless major beach preservation efforts are undertaken.

- Sea level rise would increase the costs of flooding, flood protection, and flood insurance in coastal areas. Flood damages would increase because higher water levels would provide a higher base for storm surges; erosion would increase the vulnerability to storm waves; and decreased natural and artificial drainage would increase flooding during rainstorms.
- Future sea level rise may already be an appropriate factor to consider in designing coastal drainage and flood protection structures.
- Increased salinity from sea level rise would convert cypress swamps to open water and threaten drinking water supplies.
- The adverse impacts of sea level rise could be ameliorated through anticipatory land use planning and structural design changes.
- Other impacts of global warming might offset or exacerbate the impacts of sea level rise. Increased droughts might amplify the salinity impacts of sea level rise. Increased hurricanes and increased rainfall in coastal areas could amplify flooding from sea level rise. Warmer temperatures might enable mangrove swamps—which accrete differently than salt marshes—to advance further north, perhaps changing wetland loss caused by sea level rise.
- River deltas throughout the world would be vulnerable to a rise in sea level, particularly those whose rivers are dammed or leveed.
- Economic studies of Bangladesh, Egypt, and the United States suggest that sea level rise would be economically important to coastal areas.

Notes

1. This result was reported in the North America study. The data also show it to be true in the northern Europe study, but the result was not reported. David Aubrey, Woods Hole Oceanographic Institute, Woods Hole, Massachusetts, personal communication.
2. Robert Bindschadler, Goddard Space Flight Center, Greenbelt, Maryland, personal communication.
3. James Hansen, Goddard Institute for Space Studies, New York, personal communication to J. S. Hoffman.
4. North Carolina Administrative Code, Chapter 7H, 1983. Raleigh, North Carolina: Office of Coastal Management.
5. Fred Michaud, Office of Floodplain Management, State of Maine, personal communications.

References

Alexander, C. E., M. A. Broutman, and D. W. Field. 1986. *An inventory of coastal wetlands of the USA*. Rockville, Md.: National Oceanic and Atmospheric Administration, National Ocean Service.

Ambach, W. 1985. Climatic shift of the equilibrium line—Kuhn's concept applied to the Greenland ice cap. *Annals of Glaciology* 6:76–78.

Ambach, W. (translated by G. P. Weidhaas). 1980. Increased CO_2 concentration in the atmosphere and climate change: Potential effects on the Greenland ice sheet. *Wetter und Leben* 32:135–142, Vienna. (Available at Lawrence Livermore National Laboratory, Report, UCRL-TRANS-11767, April 1982.)

Associated Press. 1985. Doubled erosion seen for Ocean City. *Washington Post*, November 14. (Maryland Section.)

Barnett, T. P. 1984. The estimation of "global" sea level change: A problem of uniqueness. *Journal of Geophysical Research* 89(C5):7980–7988.

Barth, M. C., and J. G. Titus (eds.). 1984. *Greenhouse effect and sea level rise: A challenge for this generation*. New York: Van Nostrand Reinhold.

Bentley, C. R. 1983. West Antarctic ice sheet: Diagnosis and prognosis. In *Proceedings: Carbon Dioxide Research Conference: Carbon dioxide, science, and consensus*. Conference 820970. Washington, D.C.: Department of Energy.

Bindschadler, R. A. 1985. Contribution of the Greenland ice cap to changing sea level: Present and future. In M. F. Meier et al., 1985.

Boesch, D. F. (ed.). 1982. *Proceedings of the Conference of Coastal Erosion and Wetland Modification in Louisiana: Causes, consequences, and options*. FWS-OBS-82/59. Slidell, La.: National Coastal Ecosystems Team, U.S. Fish and Wildlife Service.

Braatz, R. V., and D. G. Aubrey. (n.d.) Recent relative sea-level movement along the eastern coastline of North America. *Journal of Sedimentary Petrology*. In press.

Bruun, P. 1962. Sea level rise as a cause of shore erosion. *Journal of Waterways and Harbors Division* (ASCE) 1:116–130.

Cess, R. D., and S. D. Goldenberg. 1981. The effect of ocean heat capacity upon global warming due to increasing atmospheric carbon dioxide. *Journal of Geophysical Research* 86:498–502.

Clark, J. A., W. E. Farrell, and W. R. Peltier. 1978. Global changes in postglacial sea level: Numerical calculation. *Quarternary Research* 9:265–287.

Clark, J. A., and C. S. Lingel. 1977. Future sea-level changes due to West Antarctic ice sheet fluctuations. *Nature* 269(5625):206–209.

Davis, R. A. 1985. *Coastal sedimentary environments*. New York: Springer-Verlag.

Dean, R. G., and E. M. Maurmeyer. 1983. Models for beach profile response. In *Handbook of coastal processes and erosion*. Fort Belvoir, Va.: U.S. Army Corps of Engineers, Coastal Engineering Research Center.

Delaware River Basin Commission (DRBC). 1981. *The Delaware River Basin comprehensive (Level B) study: Final report and environmental impact statement*. West Trenton, N.J.: Delaware River Basin Commission.

Don, W. L., W. R. Farrand, and M. Ewing. 1962. Pleisstocene ice volumes and sea level lowering. *Journal of Geology* 70:206–214.

Edmonson, J., and R. Jones. 1985. *Terrebonne Parish barrier island and marsh management program*. Houma, La.: Terrebonne Parish Council (August).

Emery, K. O., and D. G. Aubrey. 1985. Glacial rebound and relative sea levels in Europe from tide-gauge records. *Tectonophysics* 120:239–255.

Emery, K. O., and E. Uchupi. 1972. Western North Atlantic Ocean Memoir 17. Tulsa, Okla.: American Association of Petroleum Geologists.

Everts, C. H. 1985. Effect of sea level rise and net sand volume change on shoreline position at Ocean City, Maryland. In *Potential impact of sea level rise on the beach at Ocean City, Maryland*. Washington, D.C.: U.S. Environmental Protection Agency.

Fairbridge, R. W., and W. S. Krebs, Jr. 1962. Sea level and the southern oscillation. *Geophysical Journal* 6:532–545.

Fastook, J. L. 1985. Ice shelves and ice streams: Three modeling experiments. In Melier et al., 1985.

Flint, R. F. 1971. *Glacial and quarternary geology*. New York: John Wiley and Sons.

Gagliano, S. M., K. J. Meyer-Arendt, and K. M. Wicker. 1981. Land loss in the Mississippi deltaic plain. In *Transactions of the 31st Annual Meeting of the Gulf Coast Association of Geological Societies*, 293–300. Corpus Christi, Texas.

Gibbs, M. 1984. Economic analysis of sea level rise: Methods and results. In Barth and Titus, 1984.

Gornitz, V. S., S. Lebedeff, and J. Hansen. 1982. Global sea level trend in the past century. *Science* 215:1611–1614.

Hands, E. G. 1980. *Prediction of shore retreat and nearshore profile adjustments to rising water levels on the Great Lakes*. Fort Belvior, Va.: U.S. Army Corps of Engineers, Tech. paper 80-7.

Hands, E. G. 1979. *Changes in rates of offshore retreat: Lake Michigan*. Fort Belvoir, Va.: U.S. Army Corps of Engineers, Coastal Engineering Research Center, publication TP79-4.

Hands, E. G. 1976. *Observations of barred coastal profiles under influence of rising water levels: Eastern Lake Michigan*. Fort Belvoir, Va.: U.S. Army Corps of Engineers, CERC publication TR76-1.

Hatton, R. S., R. D. DeLaune, and W. H. Patrick. 1983. Sedimentation, accretion and subsidence in marshes of Barataria Basin, Louisiana. *Limnol. and Oceanogr.* 28:494–502.

Haydl, N. C. 1984. *Louisiana coastal area, Louisiana: Water supply.* Initial Evaluation Study. New Orleans: U.S. Army Corps of Engineers.

Hays, J. D., and W. C. Pitman III. 1973. Lithospheric plate motion, sea level changes, and climatic and ecological consequences. *Nature* 246:18–22.

Hicks, S. D., H. A. DeBaugh, and L. H. Hickman. 1983. *Sea level variations for the United States: 1855–1980.* Rockville, Md.: U.S. Department of Commerce, NOAA-NOS.

Hoffman, J. S., D. Keyes, and J. G. Titus. 1983. *Projecting future sea level rise.* Washington, D.C.: Government Printing Office.

Hoffman, J. S., J. Wells, and J. G. Titus. 1986. Future global warming and sea level rise. In *Iceland coastal and river symposium.* ed. G. Sigbjarnarson. Reykjavik, Iceland: National Energy Authority.

Hollin, J. T., and R. G. Barry. 1979. Empirical and theoretical evidence concerning the response of the earth's ice and snow cover to a global temperature increase. *Environment International* 2:437–444.

Howard, J. D., O. H. Pilkey, and A. Kaufman. 1985. Strategy for beach preservation proposed. *Geotimes* 30(12):15–19.

Hughes, T. 1983. The stability of the West Antarctic Ice Sheet: What has happened and what will happen. In *Proceedings: Carbon Dioxide Research Conference: Carbon dioxide, science, and consensus.* Conference 820970. Washington, D.C.: Department of Energy.

Hull, C. H. J., and J. G. Titus (eds.). 1986. *Greenhouse effect, sea level rise, and salinity in the Delaware Estuary.* Washington, D.C.: Environmental Protection Agency and Delaware River Basin Commission.

Hull, C. H. J., and J. G. Titus. 1986. Responses to salinity increases. In: Hull and Titus (eds.), 1986.

Hull, C. H. J., M. L. Thatcher, and R. C. Tortoriello. 1986. Salinity in the Delaware Estuary. In Hull and Titus (eds.), 1986.

Hull, C. H. J., and R. C. Tortoriello. 1979. *Sea level trend and salinity in the Delaware Estuary. Staff report.* West Trenton, N.J.: Delaware River Basin Commission.

Kana, T. W., J. Michel, M. O. Hayes, and J. R. Jensen. 1984. The physical impact of sea level rise in the area of Charleston, South Carolina. In Barth and Titus (eds.), 1984.

Kana, T. W., B. J. Baca, and M. L. Williams. 1986. *Potential impacts of sea level rise on wetlands around Charleston, South Carolina.* Washington, D.C.: Environmental Protection Agency.

Kana, T. W., W. Eiser, B. Baca, and M. L. Williams. (n.d.). Potential impacts of sea level rise on wetlands around Little Egg Harbor, New Jersey. In *Impacts of sea level rise on coastal wetlands in the United States.* Washington, D.C.: U.S. Environmental Protection Agency.

Kennett, 1982. *Marine geology*. Englewood Cliffs, N.J.: Prentice-Hall.

Kuo, C. (ed.) (n.d.) *Potential impact of sea level rise on coastal drainage systems*. Washington, D.C.: Environmental Protection Agency.

Kyper, T., and R. Sorensen. 1985. Potential impacts of selected sea level rise scenarios on the beach and coastal works at Sea Bright, New Jersey. In *Coastal Zone, 85*, eds. O. T. Magoon et al. New York: American Society of Civil Engineers.

Kyper, T., and R. Sorensen. (n.d.) *Potential impacts of sea level rise on the beach and coastal structures at Sea Bright, New Jersey*. Washington, D.C.: Environmental Protection Agency and U.S. Army Corps of Engineers (in press).

Lacis, A., J. E. Hansen, P. Lee, T. Mitchell, and S. Lebedeff. 1981. Greenhouse effect of trace gases, 1970–1980. *Geophysical Research Letters* 81(10):1035–1038.

Laroche, T. B., and M. K. Webb. (n.d.) In *Potential impacts of sea level rise on coastal drainage systems,* ed. C. Kuo. Washington, D.C.: U.S. Environmental Protection Agency (in press).

Leatherman, S. P. 1984. Coastal geomorphic responses to sea level rise: Galveston Bay, Texas. In Barth and Titus (eds.), 1984.

Lennon, G. P., G. M. Wisniewski, and G. A. Yoshioka. 1986. Impact of increased river salinity on New Jersey aquifer. In Hull and Titus (eds.), 1986.

Lingle, C. S. 1985. A model of a polar ice stream and future sea level rise due to possible drastic collapse of the West Antarctic Ice Sheet. In Meier et al., 1985.

Magness, T. 1984. Comment. In *Greenhouse effect and sea level rise: A challenge for this generation*, eds. M. C. Barth and J. G. Titus. New York: Van Nostrand Reinhold.

Meier, M. F. 1984. Contribution of small glaciers to global sea level. *Science* 226(4681):1418–1421.

Meier, M. F., et al. 1985. *Glaciers, ice sheets, and sea level*. Washington, D.C.: National Academy Press.

Mercer, J. H. 1968. Antarctic ice and Sangamon sea level. *Geological Society of America Bulletin* 79:471.

Meyerson, A. L. 1972. Pollen and paleosalinity analyses from a Holocene tidal marsh sequence. *Marine Geology* 12:335–357.

Milliman, J. D., and R. H. Meade. 1983. World-wide delivery of river sediment to the oceans. *Journal of Geology* 91(1):1–21.

Oldale, R. 1985. Late quarternary sea level history of New England: A review of published sea level data. *Northeastern Geology* 7:192–200.

Redfield, A. C. 1967. Postglacial change in sea level in the western North Atlantic Ocean. *Science* 157:668–692.

Redfield, A. C. 1972. Development of a New England salt marsh. *Ecological Monograph* 42:201–237.

Revelle, R. 1983. Probable future changes in sea level resulting from increased atmospheric carbon dioxide. In *Changing Climate.* Washington, D.C.: National Academy Press.

Rind, D., and S. Lebedeff. 1984. *Potential climate impacts of increasing atmospheric CO$_2$ with emphasis on water availability and hydrology in the United States.* Washington D.C.: Government Printing Office.

Roberts, D., D. Davis, K. L. Meyer-Arendt, and K. M. Wicker. 1983. *Recommendations for freshwater diversion to Barataria Basin, Louisiana.* Baton Rouge: Louisiana Department of Natural Resources, Coastal Management Section.

Roemmich, D., and C. Wunsch. 1984. Apparent changes in the climatic state of the deep North Atlantic Ocean. *Nature* 307:47–450.

Roemmich, D. 1985. Sea level and thermal variability of the ocean. In M. F. Meier et al., 1985.

Schneider, S. H., and R. S. Chen. 1980. Carbon dioxide flooding: Physical factors and climatic impact. *Annual Review of Energy* 5:107–140.

Sorensen, R. M., R. N. Weisman, and G. P. Lennon. 1984. Control of erosion, inundation, and salinity intrusion. In Barth and Titus (eds.), 1984.

Stearns, L. A., and B. Mae Creary. 1975. The case of the vanishing brick dust. *Mosquito News* 17:303–304.

Thomas, R. H. 1985. Responses of the polar ice sheets to climatic warming. In Meier et al., 1985.

Thomas, R. H., T.J.O. Sanderson, and K. E. Rose. 1979. Effect of climatic warming on the West Antarctic Ice Sheet. *Nature* 227:355–358.

Titus, J. G., M. C. Barth, J. S. Hoffman, M. Gibbs, and M. Kenney. 1984. An overview of the causes and effects of sea level rise. In Barth and Titus (eds.), 1984.

Titus, J. G. 1984. Planning for sea level rise before and after a coastal disaster. In Barth and Titus (eds.), 1984.

Titus, J. G. 1985. Sea level rise and the Maryland coast. In *Potential impacts of sea level rise on the beach at Ocean City, Maryland.* Washington, D.C.: U.S. Environmental Protection Agency.

Titus, J. G. (n.d.) Greenhouse effect, sea level rise, and society's response. In *Sea Surface Studies,* ed. R. J. Devoy. Cork, Ireland: University College.

Titus, J. G., T. Henderson, and J. M. Teal. 1984. Sea level rise and wetlands loss in the United States. *National Wetlands Newsletters* 6:4.

Titus, J. G., ed. 1986. *Effects of Changes in Stratospheric Ozone and Global Climate,* Volume 4. Washington, D.C.: U.S. Environmental Protection Agency.

Titus, J. G. (n.d.) Greenhouse effect, sea level rise, and wetland loss. In: *Impacts of sea level rise on coastal wetlands in the United States.* Washington, D.C.: U.S. Environmental Protection Agency (in press).

U.S. Army Corps of Engineers. 1982. *Louisiana coastal area, Louisiana: Freshwater diversion to Barataria and Breton Sound Basins.* New Orleans: Corps of Engineers.

U.S. Army Corps of Engineers. 1980. *Feasibility report and final environmental impact statement: Atlantic coast of Maryland and Assateague Island, Virginia.* Baltimore: Corps of Engineers.

U.S. Army Corps of Engineers. 1979. *Ocean beach study: Feasibility report.* San Francisco: Corps of Engineers.

U.S. Army Corps of Engineers. 1977. *Shore protection manual.* Fort Belvoir, Va.: Corps of Engineers, Coastal Engineering Research Center.

U.S. Army Corps of Engineers. 1971. *National shoreline study.* Washington, D.C.: Corps of Engineers.

Van Beek, J. L., D. Roberts, D. Davis, D. Sabins, and S. M. Gagliano. 1982. *Recommendations for freshwater diversion to Louisiana estuaries east of the Mississippi River.* Baton Rouge: Louisiana Department of Natural Resources, Coastal Management Section.

Waddell, J. O., and R. A. Blaylock. (n.d.) *Impact of sea level rise on Gap Creek watershed in the Fort Walton Beach, Florida area.* In Kuo, C. Y. (ed.), in press.

Ward, L. G., and D. D. Domeraki. 1978. The stratigraphic significance of back-barrier tidal channel migration. *Geol. Soc. Amer.* 10(4):201.

Weishar, L. L., and W. L. Wood. 1983. An evaluation of offshore and beach changes on a tideless coast. *Journal of Sedimentary Petrology* 53(3):847–858.

Wendland, W. M. 1977. Tropical storm frequencies related to sea surface temperatures. *Journal of Applied Meteorology* 16:480.

Chapter 13

GREENHOUSE SURPRISES

Wallace S. Broecker

LESSONS FROM THE PAST

The inhabitants of planet Earth are quietly conducting a gigantic environmental experiment. So vast and so sweeping will be its impacts that, were it brought before any responsible council for approval, it would be firmly rejected as having potentially dangerous consequences. Yet the experiment goes on with no significant interference from any jurisdiction or nation. The experiment in question is the release of CO_2 and other so-called greenhouse gases to the atmosphere. As these releases are largely by-products of energy and food production, we have little choice but to let the experiment continue. We can perhaps slow its pace by eliminating frivolous production and by making more efficient our

Testimony given in a joint hearing before the Subcommittees on Environmental Protection and Hazardous Wastes and Toxic Substances of the Committee on Environment and Public Works, U.S. Senate, One-hundredth Congress, first session, 28 January 1987. The original publication of this paper provided an illustration that has not been included here. This illustration mapped sites where the Younger Dryas is found in pollen records and sites where it is not.

use of fossil fuel energy. However, beyond this we can only prepare ourselves to cope with the impacts the greenhouse buildup will bring.

As scientists, our task is to predict the consequences of this buildup. To be useful these predictions must be reasonably detailed. It is my feeling that we are in no better a position to make these predictions than are medical scientists when asked when and where cancer will strike a specific person. Understanding the operation of the joint hydrosphere, atmosphere, biosphere, cryosphere system is every bit as difficult as understanding the factors which determine whether or not cancerous cells will get the upper hand. Because of our lack of basic knowledge, the range of possibility for the greenhouse impacts remains large. It is for this reason that the experiment is a dangerous one. We play Russian roulette with climate hoping that an impact scenario of minimum consequence lies in the active chamber. While I don't claim to know what lies in the chamber, I am less optimistic about its contents than many. I suspect we have been lulled into complacency by model simulations which suggest a gradual warming over a period of 100 or so years. If this seemingly logical response to a gradual greenhouse buildup is correct, then one can imagine that man, aided by his ingenuity and technology, will be able to cope with the coming changes. While I do not have any complaints about how these modeling experiments were conducted—indeed they were done by brilliant scientists using state-of-the-art computer power—I am convinced that the basic architecture of these models denies the possibility for key interactions which occur in the real system. The reason is that at present no meaningful way exists to incorporate these interactions into models.

My impressions are more than educated hunches. They come from viewing the results of experiments nature has conducted on her own. The results of the most recent of these experiments are well portrayed in polar ice, in ocean sediment, and in bog mucks. What these records tell me is that Earth's climate does not respond in a smooth and gradual way; rather it responds in sharp jumps. These jumps appear to involve large-scale reorganizations of the Earth system. If this reading of the natural record is correct, then we must consider the possibility that the major responses of the system to our greenhouse provocation will come in jumps whose timing and magnitude are unpredictable. Coping with this type of change is clearly a far more serious matter than coping with a gradual warming.

While I cannot in this brief testimony provide a complete case in defense of my hypothesis, I can give enough of a glimpse at the evidence to capture your attention.

For over 100 years scientists have been aware that the earth is in the midst of a series of cyclic glaciations. During each of these glaciations, ice caps covered much of North America and Europe to a depth of 2 or so miles. Average global temperatures were down about 5°C. Sea level dropped over 100 meters. The CO_2 content of the atmosphere was about two-thirds its value during the warm intervals separating these icy episodes. The radiocarbon dating method tells us that the planet emerged from the most recent of these glaciations about 10,000 years ago. Mankind responded to this ease of conditions by rapidly developing a civilization. Early agriculture, for example, goes back 9000 years.

Deep sea sediments provide continuous records of these events show-

FIGURE 13.1

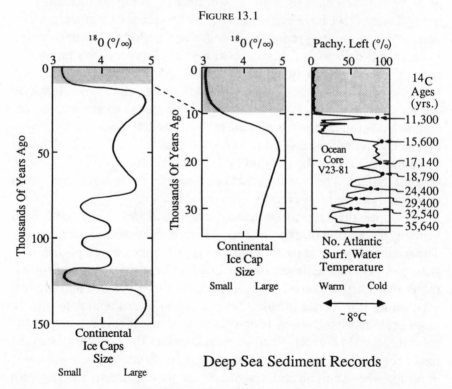

Deep Sea Sediment Records

The ratio of heavy oxygen (^{18}O) to light oxygen (^{16}O) in the foraminifera shells preserved in deep-sea sediments provides our best record as to how the great continental ice sheets of the Northern Hemisphere waxed and waned. The reason is that the snow that builds ice caps is depleted in the heavy oxygen

ing that cycles averaging 100,000 years in length carried us from warm climates, comparable to today's, to the cold climates of full glacial time. The timing of these changes points a finger at cycles in the earth's orbital characteristics as the cause for the repetitive swings. These cycles in orbital character cause the winter to summer contrast in the amount of sunlight reaching any given place on the earth to change. Seasonality was at times stronger and at times weaker than today's. While most scientists now accept the orbital hypothesis, none of the mechanisms which have been proposed to explain the linkage between seasonality and climate has received broad acceptance. This matter is the subject of considerable current research and debate.

Because the deep sea sediments we study are stirred by worms (just as the soils in our gardens) they leave the impression of smooth and gradual climate change (see Figure 13.1). Only recently have scientists

isotope. Hence, as ice caps grow, the ocean is correspondingly enriched in heavy oxygen. So also is the oxygen in the $CaCO_3$ of shells grown in these waters. The oxygen isotope record on the left covers just over one major climate cycle. There have been eight such cycles in the last million years. The shaded intervals are periods of interglaciation. The diagram in the middle is a blowup of that on the left designed to emphasize that these oxygen isotope records suggest a gradual and smooth transition from full glacial conditions about 16,000 years ago to full interglacial conditions about 8,000 years ago. Recent findings demonstrate that this view of natural climate change is misleading. The sluggish retreat of the ice and blurring of the record by burrowing worms combine to give what proves to be the wrong impression.

At the right is the record from a deep-sea sediment obtained west of the British Isles. Here the blurring by burrowers is minimal. Also, the record is based on the abundance of the shells of a planktonic herbivore named Pachy-derma (left coiling) which now lives in waters adjacent to Greenland. It is the only species found in these icy waters. During cold periods this species dominated the entire northern Atlantic, providing up to 90% of the shells found in the sediments off England. As can be seen, this record tells a different story. The northern Atlantic Ocean warmed abruptly about 15,000 years ago, and has so remained (except for two brief cold relapses) until the present. It is the abruptness of the warming and the two brief and intense epochs of renewed cold which are of great interest. The second of these cold epochs was first recognized by scientists studying the pollen records from bogs created in northern Europe during the retreat of the last great ice sheet. These records show that the forests which repopulated this area after ice departed were destroyed by a brief and intensely cold interval of several hundred years duration. They were replaced by the shrubs of glacial time. This awesome cold event is referred to by palynologists as the Younger Dryas.

begun to realize that this impression is a false one. Rather, at least in some regions of the earth (in particular the one inhabited by our NATO allies) the climate changes were abrupt. One of the early clues came from studies of the shells of microscopic organisms contained in deep sea cores of high accumulation rate obtained in the northern Atlantic. Rather than showing the gradual change scientists had become accustomed to, they showed an abrupt end to glacial time and even more interesting a brief period of intense cold interrupting the warm period which followed (see Figure 13.1).

It took more than this, however, to make scientists think seriously about the consequences of these abrupt changes. The evidence which turned our heads came from borings made through the Greenland ice cap. As no worms live in the ice and as a foot or so of ice forms from each year's snowfall, the record accurately captures the changes in the ice cap environment. These changes are recorded in the ratio of isotopically "heavy" to isotopically "light" water in the ice (a measure of air temperature), in the content of particulate matter in the ice (a measure of the dustiness of the air over the ice cap), and in the content of CO_2 and other greenhouse gases contained in the air trapped as bubbles in the ice (a measure of the atmosphere's greenhouse capacity). Like the ecologic record in deep sea muds from the northern Atlantic, the ice core records from Greenland give a dramatically different impression of the manner in which climate changes.

The ice records show climate changing frequently and in great leaps. [1] The typical leap involves a 6°C change in air temperature, a fivefold change in atmospheric dust content, and a 20% change in the CO_2 content of air. In cold times the air was dustier and had less CO_2 (see Figure 13.2). While the record in Greenland's ice tells us that climate can change in big leaps, we have to look further for a clue as to *why* these leaps occur. The last of the events seen in the Greenland record (called the Younger Dryas) is well documented not only in Greenland ice and in North Atlantic sediment, but also in lake and bog sediments from throughout western Europe and maritime Canada. It is recorded by major shifts in the ecology of the plants inhabiting the regions around these water bodies (recorded by their pollen grains). During warm periods trees grew in these areas; during cold periods the trees were replaced by tundra shrubs.

The clue we needed comes from the geographic distribution of the

FIGURE 13.2

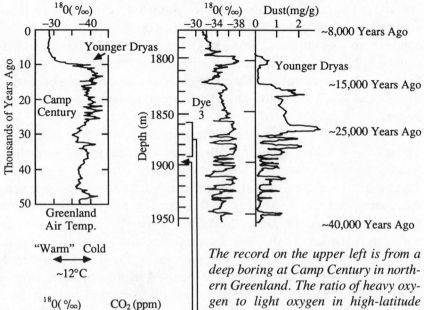

The record on the upper left is from a deep boring at Camp Century in northern Greenland. The ratio of heavy oxygen to light oxygen in high-latitude precipitation is closely tied to air temperature. The lower the air temperature, the less heavy oxygen (i.e., the more negative the $\delta^{18}O$ value). A dramatic warming ($\sim 12°$) occurred just after the Younger Dryas event close to 10,000 years ago. During the 15,000-year period of intense glaciation of the Northern Hemisphere which preceded the Younger Dryas event, the air temperature over Greenland was subject to numerous large and rapid fluctuations. By contrast, the air temperature during postglacial time has remained remarkably close to the present-day value.

The other two diagrams show oxygen isotope, dust, and CO_2 records from a deep boring at the Dye 3 station in southern Greenland. Only the record from glacial time is shown here (i.e., from about 8,000 years ago back to about 40,000 years ago). Note that each of the sharp coolings in air temperature is matched by a large increase in the dust content of the ice. Where CO_2 has been measured in the air of bubbles trapped in the ice, these sharp coolings are matched by sudden drops in CO_2 content. The oxygen isotope and dust records were obtained by Dansgaard and his group in Copenhagen, Denmark. The CO_2 records were obtained by Oeschger and his group in Bern, Switzerland. The Greenland ice cores were obtained by a team made up of Americans, Danes, and Swiss.

Younger Dryas event provided by this pollen evidence. While the event is found in bog and lake sediments throughout northern Europe, it is not seen in similar records from the United States. Rather, its impacts are confined to the northern Atlantic Ocean and its surrounding landmasses. This geographic distribution points at the northern Atlantic Ocean as the causal culprit. Based on this clue, scientists are beginning to see how devilish are the links between components of the climate system.

To understand this, we must delve into the ocean's role in climate. Those of us who live in coastal cities know that heat stored in offshore waters during the summer is released during the winter, greatly ameliorating the intensity of seasonal change (compared to that experienced by the continental interior). Less well known is the fact that western Europe is warmed by heat released from the surface waters of the northern Atlantic. In this case the amount of heat released is so great that it cannot be accounted for by storage during the summer months. It amounts to a staggering 30% of that received by the northern Atlantic from the sun! Rather, this heat is steadily carried northward by a conveyor-belt-like ocean circulation system (see Figure 13.3). This conveyor belt carries more than ten times the combined flow of the world's rivers! This water moves northward into the region around Iceland. Here it gives up the heat it gained at lower latitudes to the winter winds blown in from northern Canada. The cooling makes the water more dense, causing it to sink to the abyss. From here it flows through the deep sea as a great river, down the full length of the Atlantic, around Africa through the southern Indian Ocean, and finally up the Pacific Ocean (see Figure 13.3).

It is important to note that in the North Pacific the ocean conveyor belt runs just the other way round from that in the Atlantic. Deep waters move toward the north and upwell to the surface and then move equatorward in the upper ocean. So in today's world the Atlantic Ocean conveyor belt carries tropical heat for delivery to the atmosphere at high northern latitudes, while the Pacific conveyor belt forces cold surface waters to move southward, pushing the invading warm waters back toward the equator.

Why does our ocean operate in this fashion? While we don't have the complete answer, we do have the first principles. The circulation system is governed by salt. Because of the difference in the circulation pat-

FIGURE 13.3

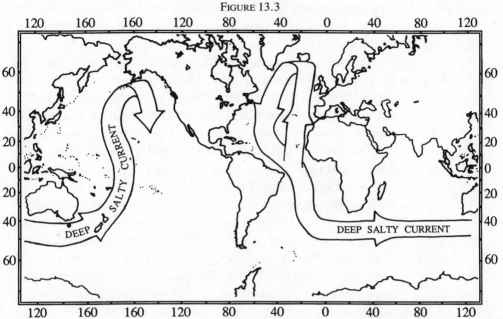

A large-scale salt transport system operates in today's ocean as a counterbalance to the transport of water (as vapor) through the atmosphere from the Atlantic to the Pacific Ocean. Upper waters in the North Atlantic flow northward to the region around Iceland. Here they give up heat to the atmosphere and descend to the abyss as new deep water. This deep water flows down the length of the Atlantic around Africa through the southern Indian Ocean and finally northward in the deep Pacific Ocean. Some salty water upwells in the northern Pacific bringing with it the salt left behind. This atmospheric water vapor and ocean salt transport system is self-stabilizing. Records from ice and sediment tell us that this great conveyor system was somehow disrupted during glacial time and replaced by an alternative mode of operation.

terns, surface waters in the North Atlantic are on the average warmer than those in the North Pacific. This allows more water to evaporate from the North Atlantic than from the North Pacific and in turn gives rise to a net transport of water vapor through the atmosphere from the Atlantic to the Pacific. As the ocean's salt can't evaporate, the North Atlantic is enriched in salt by this process (and the North Pacific waters correspondingly diluted). The enrichment of salt in the North Atlantic must somehow be compensated by a flow of salt through the sea from Atlantic to Pacific. The compensation in today's ocean is achieved by the flow of a deep current of salty water from the Atlantic to the Pacific and a matching flow of corresponding less salty water around the other way in the upper ocean.

The phenomenon which maintains this situation is a devilish one and I suspect a potentially dangerous one. The circulation pattern is self-reinforcing and hence self-stabilizing. The deep current is driven by the extra density supplied to the northern Atlantic by the enrichment of salt. The enrichment of salt is driven by the heat carried by the warm water which flows northward in the Atlantic to supply the deep current. A classic chicken and egg situation! Excess evaporation causes the deep current; the deep current causes excess evaporation.

Now back to the sharp events. The evidence we have points to a shutdown of the North Atlantic conveyor belt circulation during glacial time. Such a shutdown would cool the northern Atlantic and its adjacent lands by 6 to 10°C. This in turn would cause the boreal forests in these areas to give way to tundra shrubs. The sparsity of vegetation would permit far more dust to be lifted into the atmosphere. Finally, the only feasible mechanisms scientists have come up with to explain rapid changes in the atmosphere's CO_2 content recorded in ice cores involve modifications in the ocean's circulation pattern and intensity. Thus we surmise that the reorganization of ocean circulation which accompanied the shutdown of deep-water production in the northern Atlantic also produced the change in the CO_2 content of the earth's atmosphere recorded in ice cores. The many leaps in climate seen in the glacial part of the Greenland records very likely represent flips of the system back and forth between two self-stabilizing modes of operation.

The important point of all this is that the great leaps in the climate of the northern Atlantic region have opened our eyes to an entire new realm of climate interactions. No one will deny that the pattern and rate of global-scale ocean currents are strongly tied to the alterations in salt distribution caused by the transport of water vapor through the atmosphere from one ocean basin to another. No one will deny that the temperature of high-latitude landmasses is strongly influenced by heat transported by currents in the adjacent ocean. No one will deny that the content of CO_2 in the atmosphere depends on a linkage between ocean circulation and ocean life cycles. Further, most would agree that the present mode of operation is not God-given. The climatic records kept in ice and sediment speak strongly to us that past climate changes have involved such reorganizations. If so, why not the future climate changes as well? The increase in the atmosphere's burden of greenhouse gases may force a reorganization of the Earth system into yet another mode.

At this point it must be emphasized that the leaps in climate seen in the ice and sediment records for the northern Atlantic region were confined to the cold parts of the climate cycle. Following the transition from cold to warm conditions 10,000 years ago, the climate in this region has remained remarkably constant. Apparently the Earth system has been firmly locked in its present mode of operation for this entire period. If so, what is the likelihood that increases in CO_2 and other greenhouse gases will be able to jolt the ocean–atmosphere system out of its current mode into one more suitable to the coming conditions?

Unfortunately, we have little basis for answering this important question. Nature provides us with no recent analog of the superinterglacial conditions we are about to generate. The climate of the last 10,000 years is representative of the warmest part of the last several glacial cycles (see Figure 13.1). Hence nature has not explored the superinterglacial climate regime (at least not in sufficiently recent times that our geologic records give a rich enough picture of earth's environment to be useful for future prediction). Further, as we have only recently become aware of the complexity of the linkages which tie together the ocean, atmosphere, ice, and terrestrial vegetation, we have not even begun to formulate means by which these linkages might be modeled. Indeed, reliable prediction may never be possible. If this is the case, then our grand experiment could certainly prove to be a dangerous one!

PROPOSED ACTION

What should we do? Clearly burying our heads in the sand as we have been prone to do is irresponsible. Wringing our hands in anguish over our inability to stop the potentially dangerous experiment accomplishes nothing. There are two things we can do. First, we must explore every economically and environmentally feasible means to hold down production of greenhouse gases. Second, we must greatly expand our efforts to better understand the operation of the earth's complex environment system.

As my expertise applies only to the second of these needed actions, I will confine my remarks to it. First of all if we are to get ourselves in a position to understand the impact not only of the greenhouse gases but also of the acids and the poisons being released to the environment we must greatly expand our effort to understand individual units of the great Earth surface system and the interactions among these units. Over the next decade or so, we must ramp up our efforts by tenfold. The task is every bit as complex as the crusade to prevent cancer and the crusade to defend against nuclear missiles.

Also needed is a complete reorganization of our country's approach to environmental research. I believe that most scientists would agree with me that the handling of research on greenhouse gases by DOE and on acid rain by EPA has been a disaster. Not only do these agencies misspend much of the money allocated for such work, but because they are the designated lead agencies it is often impossible to get money from alternate sources to work on problems in which the lead agency has no interest. The best scientists are disenchanted by the necessity to deal with managers who simply don't understand the subtleties of the problems they are empowered to solve. The responsibility for basic research must be wrested from these agencies and placed in an organization which is dedicated to basic research on the environment and isolated from immediate political pressures.

I am not suggesting that this would be accomplished by simply increasing the NSF budget. NSF is not set up to put strings on the use of its funds. Nor should it be. To do the kind of environmental research which will be required if we are to effectively deal with long-term environmental problems, there must be some strings. I think that scientists working on the ocean and atmosphere have learned the value of coordination in working toward goals which are unachievable by the efforts of individual research groups or even by individual nations.

In addition to more resources and better management, we need many more properly trained scientists in this area. Such a cadre is best produced by a carefully sequenced program of graduate fellowships, postdoctoral fellowships, and career research positions.

While in this decade of budget deficits it may be unrealistic to call for new government entities, I am convinced that the only sure road to straightening out the mess which exists in the area of environmental research is to create a national institute dedicated to this endeavor. This institute would have a twofold objective. First, it would provide funding and coordination of environmental training and research in universities. Second, it would provide advice to the government on a broad range of environmental questions.

I emphasize university involvement for it is in this setting that excellent research is generated for a minimum cost. By housing graduate education, postdoctoral training, and research under one roof, an optimum environment for both training and research is created. The housing of environmental research in large laboratories operated by government agencies has, with a few outstanding exceptions, proven to be nonproductive and wasteful.

There are, however, two kinds of efforts which require facilities not available to universities. One of these is satellite observations which will be the cornerstone of our effort to monitor the system we seek to understand. Such platforms are best operated by NASA (or NOAA). The other is the very large climate models which are now the basis for all climate prediction. These facilities require large research staffs and megacomputers. Fortunately, a successful format has already been found. NASA bases such an effort on Columbia's campus. NOAA locates its primary effort on Princeton's campus. NSF has established an institute for joint university research at the National Center for Atmospheric Research in Boulder. Because of the enormous importance of these so-called "general circulation models" to climate prediction, these existing labs must be strengthened. Also, similar efforts must be launched in university settings.

As far as specific areas in which our efforts must be greatly expanded, the five listed below can be singled out for special attention:

1. The large-scale circulation of the ocean.
2. The processes regulating soil moisture.
3. The processes responsible for cloud formation.
4. The influence of global biogeochemical cycles on the atmosphere's composition.
5. The processes regulating sea ice.

Of course, programs already exist in all these areas, but in my estimation they will not bring the desired answers on the appropriate time scale. In each area we need major new observational programs to supply key data needed to develop a better physical understanding. In each area we need a cadre of young scientists with the appropriate training.

We also need to intensify our study of climate changes which have taken place over the last 100,000 years. As discussed above, nature on her own has conducted climate experiments of large magnitude. The response of the system to these natural experiments is recorded in sediments and in ice. By thoughtful study of these records, we will be able to learn valuable things about the interactions which link the various major elements of the Earth system. I would like to stress once again that it is changes in these linkages which are likely to carry the greatest threats.

SUMMARY

While we don't know nearly enough about the operation of the earth's climate system to make reliable predictions of the consequences of the greenhouse buildup, we do know enough to say that these impacts are potentially quite serious. Regardless of the scenario adopted, it seems to me that the earth's remaining wildlife will be dealt a serious blow. If, as the climate record in ice and sediment suggests, climate changes come in leaps rather than gradually, then the greenhouse buildup may threaten the continuity of our food supply. To date, we have dealt with this problem as if its impacts would come in the sufficiently distant future and so gradually that we could easily cope with them. While this is certainly a possibility, there is, I believe, an equally large possibility that the impacts will be considerably more dramatic. Hence, we must pull our heads out of the sand and deal with the climate problem as with the cancer problem and nuclear defense problem. As there are no quick fixes or easy solutions, we must gear up for the long, hard job of figuring out how the Earth system operates. To do this will require not only more financial and human resources, it will also require a management scheme appropriate to the task. Not only do our current managers not have a proper intellectual grasp of the problem, but they are obsessed with legislatively imposed "five year reports" and give little attention to developing a long-term strategy to build the needed base of knowledge. Even with a great intensification of effort, I fear that the greenhouse impacts may come largely as surprises. However, the greater our knowledge, the greater the wisdom that will be brought to bear if surprises do come.

Note

1. *Editor's Note:* Following the hearing, Senator Mitchell posed the following question, to which Broecker responded:

   Question: You hypothesize that the earth may experience climate changes so sudden and dramatic that we will have no opportunity to adapt or anticipate. In what time-frame might these sharp jumps occur? Should we be preparing ourselves in terms of the next few decades or centuries?

   Answer: Unfortunately, the record of climate over the last several hundred thousand years provides no analog to the warm climates to be created by the greenhouse buildup. Thus, I cannot say whether the jumps in climate seen during the transition between glacial and interglacial time will characterize transition to the coming superinterglacial. However, if the climate system does change mainly through reorganization, then it would be fair to say that a jump should occur sometime during the next 100 years. Our preparation for such an eventuality must be a continuing effort. As with cancer, we are not likely to gain the answers we seek in the next 50 years.

PART IV

THE GREENHOUSE GASES

Chapter 14

SOURCES, SINKS, TRENDS, AND OPPORTUNITIES

Peter Ciborowski

Each year the emission of greenhouse gases commits the earth to a warming of 0.02 to 0.06°C. Many of these gases are released as by-products of fossil fuel combustion. The remainder are produced as a result of forest clearing in the tropics or agriculture or industrial activities. Carbon dioxide (CO_2) is the most important greenhouse gas, contributing about half of global heating. In addition there are what are known as the non-CO_2 greenhouse gases: methane (CH_4), nitrous oxide (N_2O), freon CFC-12 (CF_2Cl_2), freon CFC-11 (CF_3Cl), and tropospheric ozone (O_3). Carbon monoxide and the nitrogen oxides, while not themselves greenhouse gases, increase the amount of methane and ozone in the troposphere. There are also about 15 or 20 other greenhouse gases of lesser importance.

The significance of the non-CO_2 greenhouse gases to future warming depends on their absorption of infrared radiation. The atmosphere is nearly transparent in a certain region of the infrared—the so-called atmospheric window (see Chapter 9). About one-fifth to one-third of the heat that the planet loses to space is radiated at wavelengths in this window. If a gas absorbs in the window region, it is capable of perturbing global climate even at atmospheric concentrations measured in parts per trillion (10^{-12}) or parts per billion (10^{-9}). Many of the non-CO_2 trace gases have absorption bands either in this atmospheric window (CFC-12, CFC-11) or in the wings of this window (CH_4, N_2O).

The atmospheric concentrations of the greenhouse gases are increasing because their release rates exceed the rate of removal by natural processes. The greenhouse gases, their preindustrial concentrations, their present concentrations, their sources, and their growth rates are summarized in the tables of Chapters 15 and 16.

CARBON DIOXIDE

The atmospheric concentration of CO_2 has increased nearly 30% within the last 100 years. Carbon dioxide is released to the atmosphere as a result of the combustion of fossil fuel, deforestation, and as a consequence of global warming itself. Roughly 4 metric tons of CO_2 is produced per ton of carbon that is burned. Fossil fuel combustion and forest clearing have added at least 170 billion metric tons of carbon to the atmosphere since 1850.

Between 5.5 and 6 billion metric tons of carbon is now being released to the atmosphere per year as a result of fossil fuel use. The amount of carbon being transferred from the biota to the atmosphere as a result of deforestation is a matter of dispute. Estimates vary between 0.4 to 1.6 billion metric tons[1] and as much as 3 billion metric tons per year (Chapter 5). The atmospheric concentration of CO_2 is now nearly 350 ppm and rising by 1.5 ppm (0.4%) per year.

Coal and oil combustion each account for about 40% of the total fossil fuel emission and natural gas about 20%. For an equal amount of net energy, coal releases about 25% more CO_2 than oil and about 75% more than natural gas. The quantities of CO_2 produced from all energy sources have been detailed by MacDonald.[2]

Global oil and gas resources are limited and will be depleted sometime in the next 100 years. About 150 billion metric tons of carbon remains in recoverable oil deposits and between 100 and 150 billion metric tons in conventional gas deposits. The present atmospheric level of CO_2 is nearly 350 ppm. If all remaining global oil and gas resources were to be burned, this level would rise by about 50 ppm. Between 100 and 200 billion metric tons of carbon is thought to remain in tropical forest ecosystems—an amount roughly equivalent to the carbon content of remaining recoverable oil and gas. Recoverable global coal resources contain at least 3,500 or 4,000 billion metric tons of carbon. This is more than enough to triple the amount of CO_2 in the atmosphere.

The biota and soils in the higher latitudes are quite sensitive to a global warming. As the mean global temperature rises, soil and plant respiration rates increase, perhaps by more than 50% (see Chapter 5). This could result in the transfer from biotic reservoirs to the atmosphere of 50 to several hundred billion metric tons of carbon—an amount roughly equal to the carbon remaining in global oil and natural gas resources.

The present CO_2 releases from fossil fuel use are known with greater accuracy than are the releases from deforestation and from the increased respiration which has accompanied global warming. We do know, however, that approximately 3 billion metric tons of carbon is accumulating in the atmosphere each year. The remainder—the difference between total carbon releases and the 3 billion metric tons retained in the atmosphere—is being transferred from the atmosphere into the oceans. Even if emissions were to cease today, atmospheric CO_2 would remain elevated for at least 300 years.[3]

METHANE

The atmospheric level of methane has approximately doubled over the last 100 years and is now increasing by about 0.016 ppm annually— 1.1% per year. Its present concentration is about 1.7 ppm. Methane began to increase in concentration in the atmosphere in the early seventeenth century, coinciding with the beginning of the global population expansion.

Methane is produced during the anaerobic decomposition of organic matter in waterlogged soils, swamps, marshes, landfills, rice paddies, and freshwater and marine sediments, from biomass burning, and as a result of leakage from natural gas pipelines and coal mines (see Chapter 17). It is also produced in the intestines of cattle and other ruminants.

With the exception of coal mining and natural gas leakage, which account for about 15% to 20% of present emissions (50–120 million metric tons), most emissions of methane are biogenic in origin. Although there is considerable uncertainty about the sources of methane, it appears that cattle and rice paddies together account for about 35% to 40% of emissions (120–240 million metric tons); landfills account for about 10% (35–60 million metric tons); and biomass burning about 10% to 15% (35–90 million metric tons). Natural sources (termites, ocean and lake sediments, and wetlands, for example) may account for about 15% to 30% of emissions.

Between 350 and 600 million metric tons of methane is released to the atmosphere annually. Most of this is removed from the atmosphere photochemically through reactions involving the hydroxyl radical (OH) which in turn is intimately linked with atmospheric carbon monoxide (CO); see Chapter 17. A small fraction of the methane escapes to the stratosphere, where it is oxidized to produce stratospheric water vapor—a powerful greenhouse gas. Approximately 60 million metric tons of methane accumulates annually in the troposphere. Methane then remains in the atmosphere for about 10 years.

In the future, changes in atmospheric methane will depend both on emissions and on how the atmospheric sink for methane evolves. It is well established that if emissions of carbon monoxide were to increase significantly, the amount of CH_4 that could be removed from the troposphere annually through oxidation would decline. As a result, more CH_4 would accumulate in the atmosphere and the level of methane in the global atmosphere would rise. Thus while carbon monoxide is not itself a greenhouse gas, its presence in the atmosphere adds to greenhouse warming. Indeed, about one-third of the observed increase in methane is likely to have resulted from past emissions of carbon monoxide from fossil fuel combustion. A doubling of carbon monoxide emissions would increase the atmospheric abundance of methane by about one-third.

CARBON MONOXIDE

Carbon monoxide is formed through incomplete oxidation of carbon. It has a 5-month residence time in the atmosphere, where its concentration is increasing from 0.8% to 1.5% per year.[4] About three-quarters of the present global emission of 1 billion metric tons of carbon as carbon monoxide results from human activity. Forest clearing in the tropics accounts for about 15% (160 million metric tons), and the burning of savanna for agricultural purposes adds another 10%. Fossil fuel burning produces about a quarter of the total CO emission (230 million metric tons), about half of which is released by automobiles and trucks.[5] Finally, about one-quarter of the total emitted CO results from photochemical oxidation of methane. The principal natural sources of carbon monoxide include the oceans, plant emissions, and photochemical oxidation of natural nonmethane hydrocarbons in the atmosphere (240 million metric tons).

Most carbon monoxide is removed from the atmosphere by photochemical reactions with OH. The soils also constitute a weak sink for CO. About 10 million metric tons of carbon as carbon monoxide accumulates annually in the atmosphere.[6]

TROPOSPHERIC OZONE

Most tropospheric ozone is produced photochemically in the atmosphere as a result of the oxidation of carbon monoxide, methane, or other hydrocarbons (such as butane, acetylene, propane) in the presence of nitrogen oxides (NO_x) which act as catalysts. Any substantial increase in NO_x emissions will increase the amount of ozone in the troposphere. Tropospheric ozone also increases as emissions of carbon monoxide or methane increase.

There are indications that, as a result of industrialization, tropospheric ozone levels have risen. An increase of between 20% and 50% is suggested for the past 100 years. At present, tropospheric ozone appears to be rising by about 1% per year in the Northern Hemisphere.[7] If the present trends in anthropogenic carbon monoxide emissions and the observed rate of increase in atmospheric methane were to continue, by the year 2035 tropospheric ozone concentrations could increase by a further 40% to 50%.[8] This calculation suggests that up to a doubling is possible in the next century. Alone, a doubling of methane emissions would increase tropospheric ozone about 15% in the Northern Hemisphere. Without any increase in CH_4 or CO emissions, a fourfold rise in energy-sector NO_x emissions would result in an increase in Northern Hemisphere tropospheric ozone on the order of 25%. As much as a 10%

to 20% increase would result from a doubling of anthropogenic carbon monoxide emissions.[9]

Tropospheric ozone is removed from the atmosphere partially through photochemical reactions with OH. Thus, like methane, the rate at which ozone in the troposphere is photochemically removed depends on emissions of carbon monoxide. If CO emissions rise, as is thought likely, they will slow the rate of removal of ozone from the troposphere, lengthen its atmospheric lifetime, and further increase its concentration.

NITROGEN OXIDES

Nitrogen oxides are produced principally as a result of fossil fuel use and biomass burning. About 50 million metric tons of NO_x is produced each year. Biomass burning accounts for about one-quarter of the total; the combustion of fossil fuels accounts for about 40%.[10] The remainder is produced naturally by soils and lightning. Of the fossil component, about one-third is released through the combustion of coal and 40% is released from the global transportation sector. Global NO_x emissions increased at a rate of about 4.6% per year from 1965 to 1975. A threefold increase in NO_x emissions from the energy sector is projected for the next 50 years.[11]

NITROUS OXIDE

Atmospheric levels of nitrous oxide (N_2O) began to increase with the advent of industrialization. The sources of nitrous oxide are not well understood. Between 3 and 4 million metric tons of N_2O as nitrogen are thought to be produced annually from fossil fuel combustion, principally coal. The amount of nitrous oxide destroyed annually in the stratosphere is about 10 million metric tons. About 3.5 million metric tons accumulates annually in the atmosphere. Thus the total annual production is probably about 14 million metric tons, about 25% from fossil fuel use. The remainder results from emissions from soils, the denitrification of agricultural fertilizers, biomass burning and deforestation, and oceanic emissions. Thus it might be possible to stabilize the atmospheric level of N_2O by completely eliminating fossil fuel releases or by limiting fossil fuel releases in conjunction with limits on deforestation and fertilizer-related emissions.

Of the fossil fuel source, about half is produced in the electrical sector and half in the industrial sector. Most of this amount (about 85%) is associated with coal combustion. In fact, twice as much N_2O is released as a result of coal combustion as from oil combustion. The burning of natural gas releases about one-tenth the N_2O of coal combustion.

Releases of N_2O from soils are uncertain but may be comparable to releases from coal combustion. If releases from disturbed soils are of the same order as releases from fossil fuel combustion, it is possible that N_2O soil releases may increase as global population and food needs expand. In this case, it is not clear how significant atmospheric accumulation of N_2O could be avoided.[12]

Nitrous oxide is destroyed by ultraviolet light and has a 150-year residence time in the atmosphere. Concentrations of nitrous oxide are increasing presently at about 0.3%, or 0.7 ppb, per year. A 50% increase of the present concentration would raise the mean global temperature between 0.2 and 0.5°C.

Nitrous oxide sources exceed its sinks by as much 30%. Thus even fairly conservative projections for future world coal use (say, a 1% per year growth) would increase atmospheric levels of N_2O by up to 20% over the next half century.[13]

CHLOROFLUOROCARBONS

The chlorofluorocarbons (CFC-11, CFC-12, CFC-22, CFC-113, CFC-114, CFC-115) are synthesized chemicals that have no natural counterparts. CFC-12 is used principally in refrigerators, mobile air conditioners, aerosols, and the manufacture of polystyrene. The principal uses of CFC-11 include the manufacture of urethane foams and use in aerosol spray cans. The United States accounts for about one-quarter of the total global consumption of these two chemicals. Uses within the U.S. economy include mobile air conditioning (24%), rigid and flexible urethanes (34%), refrigeration (6%), air conditioning (3%), other uses (26%). CFC-22 is used primarily in residential air conditioners. CFC-113 is used as a solvent in the electronics industry for cleaning circuit boards and for metal degreasing. CFC-114 and CFC-115 are used in refrigeration and cooling. The atmospheric lifetimes of the chlorofluorocarbons are typically 50 to 200 years. Hence, even with a weak emission source, significant atmospheric accumulations are possible. These gases are all powerful infrared absorbers.

The chlorofluorocarbons are known to be stratospheric ozone-depleters. Once released to the atmosphere, these gases are transported to the stratosphere where they catalyze the reactions that destroy stratospheric ozone (see Chapter 15). As a result, they have been targeted for regulation under the Montreal Protocol on Substances That Deplete the Ozone Layer. Under this agreement, an emissions reduction of 15% to 30% is likely to be realized (see Chapter 20). This suggests that concentrations of CFC-11 and CFC-12, now increasing by about 5% per year, will eventually reach an aggregate level of 2 to 2.5 ppb. A substantial tightening of the Montreal Protocol would constitute an important effort to limit greenhouse emissions. An 85% reduction would be required for concentrations of the principal CFCs to stabilize.[14] CFC-22, one of the substitutes for CFC-12 in air conditioning and refrigeration, can be expected to increase significantly in concentration under the Montreal Protocol. It is growing rapidly in the atmosphere (10% to 14% per year). It has been thought that a 1-ppb increase in concentration, which at the present rate of increase might be expected in 20 to 25 years, would elevate the mean global surface temperature 0.04 to 0.15°C, but recent investigations of infrared absorption of this compound suggest an upward revision.[15] Hence in the future this compound may rival CFC-11 or CFC-12 in its climatic significance.

FULLY FLUORINATED COMPOUNDS

Fully fluorinated compounds do not dissociate in the stratosphere and hence are not regulated under the Montreal Protocol. The two principal fully fluorinated compounds for which we have information on trends are carbon tetrafluoride (CF_4) and CFC-116 (C_2F_6). CFC-116 is unknown in nature. Due to very long atmospheric residence times (10,000 years), substantial atmospheric accumulations of these gases are possible over even short periods of time. Natural sources of carbon tetrafluoride are probably small.[16]

Carbon tetrafluoride and CFC-116 are released to the atmosphere principally during the refining of aluminum. Future emissions will depend on developments in the global aluminum industry. A short-term growth of 3% per year is typically forecast for this industry.

OTHER GASES OF POTENTIAL SIGNIFICANCE

Other potential greenhouse gases include methyl chloroform, methylene chloride, Halon-1301, CFC-13, CFC-14, carbon tetrachloride, peroxyacetyl nitrate (PAN), sulfur hexafluoride, and stratospheric water vapor. Stratospheric water vapor results from the oxidation of methane. Sulfur hexafluoride is associated with the aluminum industry and, like carbon tetrafluoride, has an extremely long atmospheric lifetime. PAN is one of the oxidation products of NO_x. Its residence time is sufficiently long that if global combustion activities increase significantly, it could increase in concentration in the atmosphere. Carbon tetrachloride is used in the production of CFC-12. Methylene chloride and methyl chloroform, used as solvents in the electronics industry, represent possible substitutes for CFC-113. It is possible that, as a consequence of the Montreal Protocol, their use will expand. Halon-1301 is employed as a fire extinguisher.

Many of these gases are extremely powerful absorbers of infrared radiation. Should the atmospheric concentrations of these last four gases, along with those of CFC-13, CF_4, and CFC-116, each rise just to 1 ppbv, collectively they could warm the planet 0.6 to 2°C. These gases bear watching.

As is the case with methane or tropospheric ozone, many of the lesser greenhouse gases (methyl chloroform, methylene chloride, methyl chloride, CFC-22, carbon tetrachloride, chloroform) are removed from the atmosphere through photochemical reactions involving OH. As emissions of carbon monoxide (or methane) increase, however, the ability of the troposphere to remove these gases photochemically may decline. Thus there could be a further increase in their concentrations beyond what is now projected.

RELATIVE IMPORTANCE OF THE GREENHOUSE GASES

The mean global temperature is quite sensitive to changing atmospheric concentrations of CO_2 and the other trace greenhouse gases. An increase in the CO_2 level to 400 to 500 ppm would result in a mean global

FIGURE 14.1
Sensitivity of Mean Global Temperature to Concentrations

Gas	Change in Concentration	Warming (°C)
CO_2	100 ppmv	0.5–1.6
CH_4	doubling	0.2–0.7
N_2O	+50%	0.2–0.6
Ozone (trop.)	+50%	0.5–1.7
CFC–12	1 ppbv	0.1–0.4
CFC–11	1 ppbv	0.1–0.4
H_2O (strat.)	doubling	0.6

warming of between 0.3 and 1°C. A doubling of the amount of methane in the atmosphere would raise the mean global temperature 0.2 to 0.7°C, while a doubling of ozone in the troposphere would raise the mean global temperature 0.5 to 1.7°C (Figure 14.1).

On a per-molecule basis, CO_2 is the weakest absorber of infrared radiation among the principal greenhouse gases. Methane is about 20 times more efficient an absorber of infrared radiation per molecule than CO_2, for example, but the large amount of CO_2 that is annually added to the atmosphere compensates for the weak absorption.

Atmospheric concentrations of CO_2 and the non-CO_2 gases are increasing annually from a few parts per trillion in the case of certain chlorofluorocarbons to a few parts per million in the case of CO_2. In the aggregate, each year's release commits the earth to a mean global heating of between 0.02 and 0.06°C (see Chapter 16). Carbon dioxide accounts for about one-half of this increase and methane, the second most important greenhouse gas, about one-fifth.

Most of this warming will not be experienced for decades. Although the atmosphere warms almost instantaneously as a result of a change in concentration of these gases, the oceans warm at a significantly lower rate, thereby slowing the rate of atmospheric response. Thus the consequences of present emissions will become evident only three or four decades from now.

CONTINUATION OF PRESENT EMISSION TRENDS

Past emissions have resulted in a substantial unrealized warming. Since the beginning of the industrial age, atmospheric levels of CO_2 have increased about 30%, the atmospheric level of methane has more than doubled, and concentrations of N_2O have increased about 10%. Tropospheric ozone levels were possibly 25% lower, and there were no CFCs in the preindustrial atmosphere. As a result of these changes in atmospheric concentration, the mean global temperature will eventually rise between 0.8 and 2.4°C.[17]

Without policy intervention, future emissions are likely to follow present trends—among other things, a continuation of the present rate of expansion in coal use, an implementation of the Montreal Protocol by the industrialized nations, and a continuation of the present rate of increase in methane levels in the atmosphere. A continuation of present trends in tropical forest clearing would probably result in at least a 50% loss of the tropical forests.[18] In the aggregate, a continuation of these trends over the next 50 years would commit the earth to a rise in mean global temperature of about 1.5 to 5°C (see Figure 14.2 and Chapter 16) in addition to the warming to which we are already committed as a result of pre-1985 emissions.

These projections do not include the warming which would result from releases of greenhouse gases in response to global warming itself. As global temperature increases, temperature-sensitive emissions become increasingly important. Methane is released upon the anaerobic decay of organic matter. The rate of anaerobic respiration increases with

FIGURE 14.2
Future Concentrations and Warming: 1985–2035

Gas	2035 Concentration	Warming (°C)	Atmospheric Lifetime (years)
CO_2	450 ppmv	0.58–1.73	300–500[c]
N_2O	375 ppbv	0.09–0.26	180
CH_4	2.95 ppmv	0.21–0.64	10
CFC-12	1200–1600 pptv	0.10–0.46	130
CFC-11	700–1000 pptv	0.05–0.26	65
CFC-113	190 pptv	0.02–0.07	90
O_3 (trop.)	+50%	0.24–0.71	0.1–0.3
CFC–22[a]	1.23 ppbv	0.05–0.15	20
Ozone depletion	—	0.1 –0.2	—
H_2O (strat.)	+10–15%	0.05–0.2	2
Other[b]	—	0.03–0.09	—
Total	—	1.5–4.8	

[a]5% per year increase.
[b]CH_2Cl_2, CH_3CCl_3, $CHCl_3$, CFC-13, CFC-116, CF_4. A 3% per year growth is assumed.
[c]For the combined lifetime for the atmosphere, biosphere, and upper ocean.

increasing temperature, as discussed by George M. Woodwell (Chapter 5) and Donald R. Blake (Chapter 17). With a warming of 3 to 4°C, the tropospheric level of methane might rise 60% to 70%.[19] To this increase would also be added methane that is now buried in frozen sediments off the continental shelf in the Arctic Basin. As it warms, some of this methane will be released—an amount perhaps even larger than the respiration-related release.[20]

The rate of aerobic respiration also depends on temperature. Carbon dioxide is released through aerobic respiration, particularly from soils. Although this issue has yet to be fully addressed, it appears possible that the ambient CO_2 level could increase anywhere from 20 ppmv to something well in excess of this figure.[21] About half of the committed warming will actually be experienced within 50 years, the remainder within the subsequent half century.

OPPORTUNITIES FOR LIMITING GREENHOUSE GAS EMISSIONS

NON-CO_2 GASES

Emissions of CFC-11, CFC-12, CFC-114, and CFC-115 will be reduced under the terms of the Montreal Protocol. The effects of reducing emissions of these gases would be partially offset by increasing emissions of substitute gases, many of which are themselves greenhouse gases (see Chapter 16, Figure 16.1). With the exception of CFC-22, however, these substitutes are much less efficient greenhouse gases than CFC-11 or CFC-12. Among the substitutes now being developed for these chemicals are CFC-123, CFC-134a, CFC-142b, and CFC-22. In fact, CFC-22 is already available as a substitute for CFC-12 in many applications. The remainder are in the testing stage and may be into commercial production within 5 to 10 years. These gases have much shorter lifetimes than CFC-12, or contain no chlorine, and thus are unlikely to deplete stratospheric ozone.

Under the Montreal Protocol, the gases could be stabilized in the atmosphere near 2 ppb. It is possible that greater reductions could be made in response to heightened concern over ozone depletion. At some point, however, emissions from noncomplying industrializing nations become increasingly important (see Chapter 20).

Emissions of CFC-113 will be more difficult to limit as there are no obvious substitutes for CFC-113 in certain industrial applications, principally electronics and computer manufacture, now some of the most sought-after industries in the industrialized world. Emissions are limited under the Montreal Protocol, but the absence of available substitutes suggests that it may prove difficult to meet the negotiated limits.

It is not clear that anything can be done to limit the continuing atmospheric accumulation of methane. The driving force behind present releases are agricultural and demographic in nature—an exploding population is to blame for most releases. If emissions are to be effectively controlled, world population cannot be allowed to continue along its present trajectory—it must be stabilized, and soon. Some reductions in present emissions would result from limiting the use of fossil fuels, but it certainly will not be possible to limit releases of methane resulting from increased respiration due to rising mean global temperature.

The tropospheric level of ozone is tightly coupled with that of methane. The same is true of stratospheric water vapor. To the degree that no control appears feasible for methane, it will not be possible to control much of what happens with tropospheric ozone or stratospheric water vapor.

To the extent that nitrous oxide is released as a result of forest clearing and the use of agricultural fertilizers, it is doubtful that anything can be done—essentially for the same reasons that control of rising methane emissions is so problematic. If coal combustion proves to be the principal source of nitrous oxide, then N_2O will be controlled to the extent that CO_2 is controlled because they both come from that same source.

The other known greenhouse gases contribute substantially less to the expected global warming than do the big six—CO_2, methane, CFC-11, CFC-12, tropospheric ozone, and nitrous oxide. Since emissions are distributed among many nations, the effect of emissions control by any single nation would be minuscule. Hence control is unlikely. Further, many of these gases (methylene chloride, methyl chloroform, chloroform) are substitutes for the CFCs, which will make control doubly difficult. Carbon tetrachloride emissions should decline, however, to the extent that present releases are a result of the manufacture of chlorofluorocarbons.

CARBON DIOXIDE

Several approaches have been taken to the question of how quickly we might replace fossil fuel use in the global economy, and at what ultimate level of CO_2 buildup. Some analysts have considered the rate at which capital could be diverted into new energy systems.[22] Others have based their projections on the past experience of new energy sources penetrating the market.[23] Several studies have evaluated the extent to which increased end-use efficiency could reduce fossil fuel use and hence limit the rise in atmospheric CO_2 levels.[24] Other analysts have taken an approach based on economics, asking how quickly energy demand might respond to various tax policies or other governmental interventions into energy markets.

In the aggregate, these studies suggest that a ceiling of between 450 and 600 ppm of CO_2 is achievable over the long term—even given another doubling of the global population.[25] It will, however, require the wholesale abandonment of coal as a principal energy source. This ceiling corresponds to an increase in mean global temperature of between 0.5 and 4°C above that to which we are already committed.

To a large extent, population expansion is the driving force behind releases of CO_2 from deforestation. If global population cannot be stabilized, it seems unlikely that CO_2 releases from deforestation will decline much. Moreover, there are no obvious means to limit releases from the upper layer of the oceans as it warms, from the biosphere as a result of increasing biotic respiration in high latitudes, from cement making, or from the decay of methane in the atmosphere. The unavoidable buildup of CO_2 from these sources plus deforestation would probably be at least 30 to 50 ppm.

The unavoidable buildup from the fossil fuel sector would probably put the long-term atmospheric level at something like 450 to 600 ppm and the aggregate level with biospheric sources included slightly higher, perhaps 480 to 650 ppm.

VERY-LONG-TERM WARMING

Assuming that this reasoning is correct, unavoidable future releases of the greenhouse gases would raise the mean global temperature a minimum of 2°C. Future emissions of CO_2, even assuming that concentrations stabilize between 450 and 600 ppm, would raise the mean global temperature 0.7 to 2.4°C. If we assume that the goals of the Montreal Protocol are met, and that the rise in nitrous oxide results exclusively from emissions from the energy sector, the future atmospheric buildup

FIGURE 14.3
Degree of Possible Future Limitation on Global Warming

Gas	Long-Term Level		Long-Term Warming[a]
CH_4	4–6 ppmv		0.48–0.68
N_2O	335–600 ppbv		0.05–0.42
CFC-11, CFC-12, CFC-113[b]	2.5 ppbv		0.32
O^3 (trop.)	+40%		0.4
Other[c]	—		0.33
Subtotal		1.58–2.15	
CO_2 (fossil fuel)	450–600 ppmv		0.77–1.60
Other CO_2	(+) 30–50 ppmv		0.19–0.23
Subtotal	480–650 ppmv		0.96–1.83
Total			2.5 –4.0

[a]Calculated with a climate sensitivity of 2°C per doubled atmospheric level of CO_2.
[b]Assumes that the provisions of the 1987 Montreal Protocol can be completely satisfied.
[c]CF_4, CFC-116, CFC-22, CFC-13, $CHCl_3$, stratospheric H_2O, CH_3CCl_3, CCl_4, CH_2Cl_2, stratospheric ozone depletion.

of these two gases could be expected to add 0.28 to 1.7°C. Releases of all the non-CO_2 gases would result in a mean global temperature rise of between 1.2 and 3.6°C (Figure 14.3).

Thus the unavoidable long-term mean global warming could be as low as 2 to 3°C. It would be a substantial achievement if future perturbations to mean global temperature could be constrained to a range of values as low as this. On the other hand, a limiting value as low as 2°C assumes a climate sensitivity of 1.5°C (per doubled atmospheric levels of CO_2). But the climate could be twice to nearly four times as sensitive as this. If we are unlucky, and the climate turns out to be this sensitive, then it is unlikely that the rise in mean global temperature from past, present, and future emissions could be less than 5°C—and possible that it could be nearly twice this amount. In this case, it is not certain we can do anything to stabilize future climate.

CONCLUSIONS

Concentrations of CO_2 and non-CO_2 greenhouse gases are rapidly increasing in the atmosphere. As a result, within 50 years we will be committed to a mean global temperature rise of 1.5 to 5°C. And if no attempt is made to slow the rate of increase, we could be committed to another 1.5 to 5°C in another 40 years. Emissions prior to 1985 have committed us to a warming of about 0.9 to 2.4°C, of which we have already experienced about half a degree. The warming that we have yet to experience is the unrealized warming. This warming—0.3 to 1.9°C—is unavoidable.

There are reasons to believe that emissions of certain gases can be limited, but it is equally clear that little can be done about the emissions of others. As a result, a rise in mean global temperature of 2 to 9°C may be unavoidable as a result of present and future emissions. The wide range of values results principally from the wide range of uncertainty over the sensitivity of climate to CO_2 and the non-CO_2 trace gases. If the lower value is the correct value for the unavoidable warming, policy intervention can realize substantial benefits. If the higher value is correct, there may be little we can do to limit future global warming.

Notes

1. R. P. Detwiler and C. A. Hall, "Tropical Forests and the Global Carbon Cycle," *Science 239*, 42 (1988).
2. G. J. MacDonald, *The Long-Term Impacts of Increasing Atmospheric Carbon Dioxide Levels,* Ballinger Pub., Cambridge (1982).
3. J. A. Laurmann and J. R. Spreiter, "The Effects of Carbon Cycle Model Error in Calculating Future Atmospheric Carbon Dioxide Levels," *Climatic Change 5*, 145 (1983).
4. M. A. Khalil and R. A. Rasmussen, "Carbon Monoxide in the Earth's Atmosphere: Indications of a Global Increase," *Nature 332*, 242 (1988).
5. J. A. Logan, M. J. Prather, S. C. Wofsy, and M. B. McElroy, "Tropospheric Chemistry: A Global Perspective," *Journal of Geophysical Research 86* (1981): 7210; updated and reported in World Meteorological Organization, *Atmospheric Ozone 1985: Assessment of Our Understanding of the Processes Controlling Its Present Distribution and Change,* Global Ozone Research and Monitoring Project Report no. 16 (Geneva: WMO, 1986).
6. Logan et al.; see note 5.
7. G. Tiao et al., "A Statistical Trend Analysis of Ozonesonde Data," *Journal of Geophysical Research 91*, 13,121 (1986); see also Logan et al.
8. A. M. Thompson and R. J. Cicerone, "Possible Perturbations to Atmospheric CO, CH_4, and OH," *Journal of Geophysical Research 91*, 10,853 (1986).
9. L. B. Callis, M. Natarajan, and R. Boughner, "On the Relationship Between the Greenhouse Effect, Atmospheric Photochemistry, and Species Distribu-

tion," *Journal of Geophysical Research 88* (1983): 1401; S. Hameed, R. D. Cess, and J. S. Hogan, "Response of the Global Climate to Changes in Atmospheric Chemical Composition Due to Fossil Fuel Burning," *Journal of Geophysical Research 85* (1980): 7537; A. J. Owens, J. M. Stead, D. L. Filkin, C. Miller, and J. P. Jesson, "The Potential Effects of Increased Methane on Atmospheric Ozone," *Geophysical Research Letters 9* (1982): 1105.

10. D. J. Wuebbles and J. Edmonds, *A Primer on Greenhouse Gases,* DOE/NBB-0083 (Washington, D.C.: U.S. Department of Energy, 1988).

11. M. Kavanaugh, "Estimates of Future CO, N_2O and NO_x Emissions from Energy Combustion," *Atmospheric Environment 21* (1987): 463.

12. P. J. Crutzen, "Atmospheric Interactions—Homogenous Gas Reactions of C, N, and S Containing Compounds," in B. Bolin and R. B. Cook, *The Major Biogeochemical Cycles and Their Interactions, SCOPE 21* (Chichester, U.K.: John Wiley and Sons, 1983); W. M. Hao, S. C. Wofsy, M. B. McElroy, J. M. Beer, and M. A. Toqan, "Sources of Atmospheric Nitrous Oxide from Combustion," *Journal of Geophysical Research 92* (1987): 3098.

13. Hao et al.; see note 12.

14. J. Hoffman, "The Importance of Knowing Sooner," in J. Titus, ed., *Effects of Changes in Stratospheric Ozone and Global Climate* (Washington, D.C.: U.S. Environmental Protection Agency, 1986).

15. P. Varanasi and S. Chudamani, "Infrared Intensities of Some Chlorofluorocarbons Capable of Perturbing the Global Climate," *Journal of Geophysical Research 93* (1988): 1666.

16. P. Fabian, R. Borchers, B. C. Kruger, and S. Lal, "CF_4 and C_2F_6 in the Atmosphere," *Journal of Geophysical Research 92* (1987): 9831.

17. R. E. Dickinson and R. J. Cicerone, "Future Global Warming from Atmospheric Trace Gases," *Nature 319,* 109 (1986).

18. J. R. Trabalka, J. A. Edmonds, J. M. Reilly, R. H. Gardner, and D. E. Reichle, "Atmospheric CO_2 Projections with Globally Averaged Carbon Cycle Models," in *The Changing Carbon Cycle: A Global Analysis,* J. R. Trabalka and D. E. Reichle, eds. (New York: Springer-Verlag, 1986); G. Woodwell, "Biotic Effects on the Concentration of Atmospheric Carbon Dioxide: A Review and a Projection," in *Changing Climate,* Carbon Dioxide Assessment Committee, National Research Council (Washington, D.C.: National Academy Press, 1983).

19. S. Hameed and R. Cess, "Impact of a Global Warming on Biospheric Sources of Methane and Its Climatic Consequences," *Tellus 35B,* 1–7 (1983).

20. R. Revelle, "Methane Hydrates in Continental Slope Sediments and Increasing Atmospheric Carbon Dioxide," in *Changing Climate,* Carbon Dioxide Assessment Committee, National Research Council (Washington, D.C.: National Academy Press, 1983).

21. G. H. Kohlmeier, H. Brohl, U. Fischbach, G. Kratz, and E. O. Sire, "The

Role of the Biosphere in the Carbon Cycle and Biota Models," in *Carbon Dioxide: Current Views and Developments in Energy/Climate Research,* W. Bach, A. J. Crane, A. L. Berger, and A. Longhetto, eds. (Dordrecht, Holland: D. Reidel Publishing Co., 1983); see also G. M. Woodwell, "Biotic Effects."

22. A. M. Perry, "Carbon Dioxide Production Scenarios," in *Carbon Dioxide Review: 1982,* W. C. Clark, ed. (New York: Oxford University Press, 1982).

23. J. A. Laurmann, "Market Penetration as an Impediment to Replacement of Fossil Fuels in the CO_2 Environmental Problem," in *The Greenhouse Effect: Policy Responses,* D. E. Abrahamson and P. F. Ciborowski, eds. (Minneapolis: Center for Urban and Regional Affairs, University of Minnesota, forthcoming).

24. R. Williams, A.K.N. Reddy, T. B. Johansson, J. Goldemberg, and E. Larson, "A Global End-Use Energy Strategy," in *The Greenhouse Effect: Policy Responses,* D. E. Abrahamson and P. F. Ciborowski, eds. (Minneapolis: Center for Urban and Regional Affairs, University of Minnesota, forthcoming); D. J. Rose, M. M. Miller, and C. Agnew, *Global Energy Future and CO_2-Induced Climate Change,* MITEL 83–015 (Cambridge: Massachusetts Institute of Technology, 1983).

25. P. Ciborowski and D. Abrahamson, "Policy Responses to Climate Change: Opportunities and Constraints," in M. Meo, ed., *Proceedings of the Symposium on Climate Change in the Southern United States* (Norman: University of Oklahoma Press, 1987).

Chapter 15

GLOBAL WARMING, ACID RAIN, AND OZONE DEPLETION

Ralph Cicerone

GASES INVOLVED IN THE GREENHOUSE EFFECT

Scientists now know enough about the properties of chemicals that can be effective greenhouse gases that we can list the potential key contributors to the effect and we can dismiss many other chemicals that do not possess the right properties. In the table of data that accompanies the written text of my testimony (Figure 15.1), I summarize data on greenhouse gases that are actually piling up in the atmosphere. I will read the lines and in between the lines of that table now. You will see that the composition of the atmosphere and hence the greenhouse radiative forcing of the system are entering into uncharted territories.

Carbon dioxide now comprises 346 parts per million (ppm) of Earth's

Testimony given in a joint hearing before the Subcommittees on Environmental Protection and Hazardous Wastes and Toxic Substances of the Committee on Environment and Public Works, U.S. Senate, One-hundredth Congress, first session, 28 January 1987.

FIGURE 15.1
Greenhouse Gases

Gas	Concentration in Air		Present Rate of Increase (per year)
	Pre-Industrial	1986	
Carbon Dioxide (CO_2)	275 ppm	346 ppm	1.4 ppm (0.4%)
Methane (CH_4)	0.75 ppm	1.65 ppm	17 ppb (1%)
Fluorocarbon-12 (CCl_2F_2)	Zero	400 ppt	19 ppt (5%)
Fluorocarbon-11 (CCl_3F)	Zero	230 ppt	11 ppt (5%)
Nitrous Oxide (N_2O)	280 ppb	305 ppb	0.6 ppb (0.2%)
Ozone, Tropospheric (O_3)	15 ppb?	35 ppb	0.3 ppb? (1%)
			Northern Hemisphere Only
Other Fluorocarbons	Zero	see text	(5 to 15%)

Key greenhouse gases: present concentrations in air, rates of increase, and prein-dustrial-era concentrations. Ppm = parts per million, ppb = parts per billion, ppt = parts per trillion.

atmosphere. In the preindustrial era its concentration was about 275 ppm, according to measurements of old air that was taken from ice cores. This 25% increase during the last 100 to 150 years is almost certainly due to our burning of fossil fuels (coal, gas, oil) although part may be due to the decay of organic material following deforestation. Carbon dioxide concentrations are now higher than at any previous time in the last 40,000 years. During that epoch, and possibly for the last 300,000 years, carbon dioxide concentrations have moved between 180 ppm and 300 ppm, never as high as now. There are suspicions that carbon dioxide values were much higher than at present when Earth was considerably younger, several billion years ago, for example.

When carbon dioxide is measured one finds seasonal variations and slightly different values at different locations. Generally, the rate of increase of CO_2 is about 4% per decade; again, this rate varies from year to year. Up to now and for the next 20 years or so, these CO_2 increases account for half or more of the global human-induced greenhouse effect. Future values of atmospheric CO_2 will depend on fuel usage and type, details of exchange with the oceans, and biospheric growth and decay rates. A doubling of preindustrial CO_2 concentrations is not unlikely to occur in 100 to 150 years.

Methane concentrations have risen about 100% in the last 150 years or so, from 0.75 ppm to 1.66 ppm. The last 35 years have seen a 30% to 40% increase. There is no evidence that methane concentrations were ever as high as they are now although it is possible especially for the primitive Earth. The lower preindustrial concentrations probably represent natural, unperturbed background levels of this familiar gas. Swamps, marshes, and tundra are natural sources of methane ("swamp gas") and natural gas venting is possibly significant. (Methane is the principal component of natural gas.) Human activities are quite capable of having caused these recent methane increases and of continuing such a trend into the future. Humans increase the sources of atmospheric methane by expanding rice agriculture and herds of ruminant animals and by increased coal mining and natural gas drilling and transmission, and possibly by increased usage of landfills as garbage dumps. We may also be suppressing the atmosphere's natural ability to assimilate methane by dumping more pollutants like carbon monoxide into the air. As the climate warms we should be aware of the possibility of accelerated methane releases from tundra and from oceanic methane clathrates.

Methane concentrations are increasing at a rate of 1% per year as shown by direct measurements since 1978. Less direct information indicates that a 35% increase has occurred since 1951, a rate that is consistent with the 1978–86 trend. Further, measurements of old air that was extracted from dated ice cores show that methane concentrations have doubled in the last 150 years or so.

A global warming could liberate methane that is now stored in oceanic sediments or at depth in tundra. This storage involves clathrates, or cages of ice that trap methane and other hydrocarbons. These clathrates are less stable as temperature increases. Scientists have estimated that substantial amounts of methane could be so released, especially from ocean-shelf sediments, but the exact amounts and the timing of the release are not known. One hundred years or more in the future is a good guess.[1]

The two principal fluorocarbons that are implicated in the greenhouse effect and stratospheric ozone depletion are fluorocarbon-12 and fluorocarbon-11, CCl_2F_2 and CCl_3F, respectively. These are totally synthetic chemicals so neither of them were present in the preindustrial atmosphere. Fluorocarbon-12 now comprises 400 ppt of the atmosphere and fluorocarbon-11, 230 ppt. These concentrations sound small but both of these chemicals are such potent greenhouse gases and the chlorine atoms that are released from fluorocarbon-12 and -11 in the stratosphere have such large potential for destroying ozone that global effects are of concern. Also, their concentrations have been growing at rates of roughly 5% per year.

Future increases in atmospheric fluorocarbon concentrations are fully expected. Even if emissions do not increase further, concentrations will increase for the next several decades because of the long time needed for atmospheric destruction processes to equilibrate with a steady input. Emissions to the atmosphere will increase if industrial production and usage increase—there are economic projections that show a continued or growing demand for these chemicals (or possible substitutes) as refrigerants, aerosol propellants, foam-blowing and degreasing agents, and as cleaners or solvents. Without international regulations, the emissions of these chemicals could increase by 2% to 5% annually.

In the case of nitrous oxide (laughing gas, N_2O) we are unsure as to why it is increasing but measurements from all over the world have established that it is. The 1986 concentration of N_2O was about 305 ppb. Preindustrial nitrous oxide concentrations were about 280 ppb. Presently, N_2O concentrations are increasing at about 0.2% per year. This rate might sound small to you but it actually signals a large global perturbation in the source of N_2O. It takes 5,000,000 extra tons of N_2O annually to cause this increase; whatever the identity of this extra source it is now about 25% as big as the total of all natural sources. Two possible N_2O sources seem likely. One is the inadvertent loss of artificial nitrogen fertilizer from fertilized fields and from subsequent nitrogen cycling in food and wastes. Some N_2O escapes into the air from soils and water bodies during two processes of nitrogen cycling: nitrification and denitrification. The second candidate as an N_2O source is the combustion of nitrogen-rich fuels; the higher the nitrogen content of various fuels, the more the N_2O production, it appears. World demand for N fertilizers keeps growing. In 1950 only about 3,000,000 tons were produced. In recent years the annual total was over 50,000,000 tons. Similarly, as you know, fuel combustion increases at 2–4% per year although the mix of fuels varies and the rates track the general economy.

Tropospheric (lower atmospheric) ozone is potentially an important greenhouse gas. Measurements show that tropospheric ozone is increasing in concentration in the Northern Hemisphere (where there is more human influence), perhaps 1% per year, but the situation is not as clear as for other greenhouse gases. Ozone is nowhere near as inert or long-lived as the others; a blob of ozone-rich air can decay, not just by dilution but also by chemical reactions and contact with surfaces. With only a 100-day lifetime, tropospheric ozone is more variable from place to place and season to season so it's not possible to give one number for a global ozone concentration or rate of change. There are new indications from European scientists that background ozone concentrations at the turn of this century were well below those at present, perhaps only one-third as large. More on ozone later.

Lower atmospheric (tropospheric) ozone is increasing in concentration at the few measurement stations that have gathered good data in the last 10 to 15 years. These data are not as convincing as we would like, but they seem to show a 1% per year increase. Observations from Europe show increases during the 1950s but the proximity of pollution sources confounds the interpretation somewhat. A new, very rigorous, and scholarly reconstruction of data from ozone measurements in France near the turn of this century shows that background ozone levels were well less than today, perhaps only one-third or one-half of today's.

The strongest possibility to explain the ozone increases is that NO_x and hydrocarbon pollutants (that are entering the atmosphere in growing amounts) are reacting through known chemical pathways to produce more ozone. We don't know whether ozone production is greatest near pollution sources like cities or farther downwind in remote areas, nor how effective are NO_x and hydrocarbon emissions from aircraft, but two features of the ozone field data do implicate this general explanation. One is that ozone concentrations have increased more in summer months when photochemical ozone production is easier. Second, the absence of an ozone increase in the Southern Hemisphere, where there is less industry and combustion, is a meaningful sign.[2]

Other fluorocarbons and chlorocarbons deserve some note. Greenhouse contributions from trichloroethane and carbon tetrachloride are recognized but for other chemicals, for example fluorocarbon-113, whose usages and atmospheric concentrations are growing fast, very little attention has been paid so far, i.e., their atmospheric concentrations and rates of change are poorly known as are some of their infrared properties.

Carbon monoxide (CO) is not an important direct contributor to the greenhouse effect but it has the potential to accelerate the buildup of other greenhouse gases: methane (CH_4) and ozone (O_3) in the lower atmosphere (troposphere).

CO is an important chemical reactant in the troposphere; through atmospheric chemistry several interesting feedbacks can arise. First, CO reacts with the all-important hydroxyl radical (OH). If adequate amounts of nitrogen oxides (NO and NO_2) are present, O_3 is formed in the sequence of reactions that begins with CO + OH. Without adequate NO_x, O_3 is consumed in an alternate sequence of reactions. In both cases, atmospheric OH concentrations can be decreased by increasing CO; this occurs when OH and similar radicals combine to form chemicals such as hydrogen peroxide which are susceptible to downward transport and removal, effectively removing OH itself.

In assessing how CO affects methane and ozone, we need greatly improved data on NO_x in the atmosphere. For most of the global atmosphere, though, increasing CO will decrease OH. This would, in turn, cause methane concentrations to increase even if methane sources do not increase.

On balance, available evidence from measurements shows that CO is increasing in the global atmosphere at a rate of 0.5% to 2% per year. CO arises from the incomplete combustion of carbon-based fuels but also from some natural plant emissions; the CO increase could arise from either of these or from any decrease in atmospheric OH concentrations. A final comment—when CO is released into the atmosphere instead of carbon dioxide (CO_2) from some combustion process, one might think that this would favor a slower CO_2 buildup. Actually, CO is converted to CO_2 when CO reacts with atmospheric OH.[3]

THE GREENHOUSE EFFECT AND STRATOSPHERIC OZONE DEPLETION

An important relationship exists between the causes of the global greenhouse effect and the depletion of stratospheric ozone: the key greenhouse gases fluorocarbons-12 and -11, nitrous oxide, and methane also affect stratospheric ozone strongly. These gases have atmospheric survival times long enough to permit them to make the upward journey into the stratospheric ozone layer. In the harsh ultraviolet light there the fluorocarbons are decomposed yielding chlorine atoms that destroy ozone. Nitrous oxide decomposition yields nitric oxide that acts similarly. Methane acts to slow the attack on ozone but it also decomposes to yield molecular fragments that are involved in ozone destruction.

Interestingly, present models say that the chlorine attack on ozone is slowed somewhat if in the future nitrous oxide and methane continue to increase along with the fluorocarbons. The slower decrease of total ozone will result in a redistribution of ozone toward lower altitudes.

Carbon dioxide is not involved chemically but once in the upper atmosphere it acts to cool the air by radiating energy to space. This cooling should also act to slow the chlorine attack on ozone a little.

Another connection arises because stratospheric ozone itself can affect temperatures at the surface of the Earth. If, for example, the fluorocarbons, nitrous oxide, and methane all increase in the future there should be more ozone present in the lower stratosphere and a warming should occur.

More complicated phenomena may be occurring over the Antarctic where springtime ozone concentrations have decreased greatly in the past fifteen years or so. The same ozone-destroying chemicals just mentioned are probably at play but through more complex pathways, I believe. In any case it is very important to learn the mechanism behind the formation of this Antarctic ozone hole if we want to be able to predict whether the hole will worsen and spread north or whether it will stay over Antarctica or even disappear. Of all the gases relevant today, the fluorocarbons stand out in several ways. Their concentrations in the atmosphere are the smallest, their annual percentage rates of increase are the largest, and their sources are the simplest to understand; they are manmade. Ten years ago CCl_3F and CCl_2F_2 were used mostly as aerosol propellants and as refrigerants and the U.S. accounted for about half the total world production and usage. Now the aerosol application represents a smaller proportion, especially in the U.S., and the U.S. production and usage is a smaller part of the world's. Recently, two U.S. industrial groups proposed that international agreements be sought to limit the future growth of emissions to the atmosphere. In this way steps could be taken to contain future greenhouse and ozone-depletion effects.

ACID DEPOSITION AND GREENHOUSE GASES

The relationships between regional acid deposition and the global greenhouse effect are not as direct as for the greenhouse effect and stratospheric ozone depletion. The strong acids in precipitation, sulfuric, nitric, and formic, are formed from emissions of sulfur dioxide, nitrogen oxides, and hydrocarbon pollutants. These pollutants are generally short-lived compared to greenhouse gases. None of these pollutants are effective greenhouse gases although nitrous oxide is often formed in the same processes that produce the more common nitrogen oxides. Of the key greenhouse gases only tropospheric ozone is involved in the formation of atmospheric acids.

A strong and broad link between the two phenomena is made, however, through the chemistry of the atmosphere. Hydrocarbon and nitrogen oxide pollutants react to produce ozone. All of these plus carbon monoxide control the rates at which acids are formed by oxidation and they also set the levels of other oxidants, like OH radicals, that bear the atmosphere's ability to oxidize and decompose methane. It is possible that the buildup of methane stems partly from the effects of other pollutants on the whole system. The oxidizing power of the atmosphere is the link and at this time we do not understand very well at all the chemistry of the troposphere where much of this oxidizing power appears.

Note

1–3. Author's response to questions following his testimony.

Chapter 16

OBSERVED INCREASES IN GREENHOUSE GASES AND PREDICTED CLIMATIC CHANGES

V. Ramanathan

OBSERVED INCREASE IN THE POLLUTANTS

Since the dawn of the industrial era, several trace gases have been increasing in the atmosphere. Recent advances in chemical measurements have helped document the rate of increase of the trace gas concentrations. The concentrations continue to increase at significant rates as shown in Figure 16.1 for carbon dioxide (CO_2), methane (CH_4),

Testimony given in hearings before the Committee on Energy and Natural Resources, U.S. Senate, One-hundredth Congress, first session, on the Greenhouse Effect and Global Climate Change, 9–10 November 1987. This testimony is based largely on "Climate—Chemical Interactions and Effects of Changing Atmospheric Trace Gases," by V. Ramanathan, L. Callis, R. Cess, J. Hansen, I. Isaksen, W. Kuhn, A. Lacis, F. Luther, J. Mahlman, R. Reck, and M. Schlesinger.

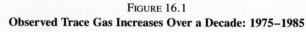

FIGURE 16.1
Observed Trace Gas Increases Over a Decade: 1975–1985

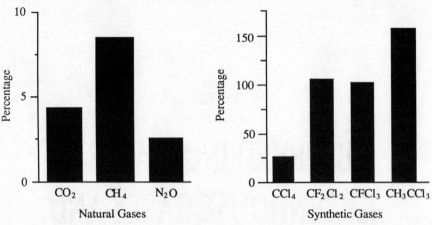

Observed trace gas increases over a decade from 1975 to 1985. CO_2 data are taken from Bolin et al., 1987, and all others are taken from Rasmussen and Khalil, 1986.

nitrous oxide (N_2O), carbon tetrachloride (CCl_4), chlorofluorocarbons (CFC), CFC-11 and CFC-12, and methyl chloroform (CH_3CCl_3). The industrial sources for these gases as well as several others that have been detected in the atmosphere are given in Figure 16.2. Studies of the chemical balance of the land–ocean–atmosphere system provide compelling arguments for attributing the observed increases to human activities such as: fossil fuel consumption, deforestation, refrigerants, propellants, and numerous others.

Scientific concern for the climatic effects of an increase in the atmosphere concentration of CO_2 dates back to the nineteenth century. It is only within the last decade that we have become aware that human activities are causing an increase in the atmospheric concentrations of not only CO_2 but several other trace gases as well.

THE GREENHOUSE EFFECT

The gases shown in Figure 16.1 trap the infrared (IR) radiation, also known as heat radiation, emitted by the surface of the earth which would have otherwise escaped to space. It is this trapping of the IR radiation which is referred to as the greenhouse effect, particularly since the gases let the solar radiation (visible light) penetrate to the surface. The synthetic trace gases are significantly more effective than others in enhancing the greenhouse effect. For example, adding one molecule of CFC-11 and CFC-12 to the atmosphere can have the same radiative effect as adding 10,000 molecules of CO_2.

A VIGOROUS CLIMATE

An increase in the concentration of the greenhouse gases leads to increased trapping of IR radiation and thus causes an excess of radiative energy available to drive the climate system. This excess energy will alter global and regional climate patterns since radiative energy (solar and IR) is the fundamental energy source for climate and the winds in the atmosphere and the oceans. The result, according to the theory, is a vigorous climate system. For example, the excess radiative energy will warm the oceans. The warmer oceans will evaporate more moisture. The excess moisture in the atmosphere will increase global rainfall. Hence, a greenhouse-rich atmosphere will be warmer, more humid, and wetter. However, because of nonlinear interactions between radiative heating and atmospheric circulation, the warming and rainfall changes will not be uniform, but will vary significantly with latitude, region, and season. In essence, shifts in the climate patterns should be anticipated.

PREDICTED CLIMATE CHANGES

- The decadal increase in the greenhouse forcing from the decade of 1850 to that of 1980 as well as from the projected increases are shown in Figure 16.3. The rate of decadal increase of the total radiative heating of the planet is now about five times greater than the mean rate for the early part of this century. Non-CO_2 trace gases in the atmosphere are now adding to the greenhouse effect by an amount comparable to the effect of CO_2 increase.
- The cumulative increase in the greenhouse forcing until 1985 has committed the planet to an equilibrium warming of about 1 to 2.5°C. The 2.5°C is based on the sensitivity of three-dimensional climate models.

FIGURE 16.2

Greenhouse Gases: Dominant Sources and Sinks, Average Atmospheric Residence Time, 1980 Global Average Mixing Ratio, and Probable Concentration Range for the Year 2030

Chemical Group	Chemical Formula	Dominant Source*	Dominant Sink*	Estimated Average Residence Time, years	Year 1980 Global Average Mixing Ratio, ppb†	Year 2030 Probable Global Average Concentration, ppb		Remarks§
						Probable Value	Possible Range	
Carbon dioxide	CO_2	N, A	O	2	339×10^3	450×10^3	350–450	Based on a 2.4% per year increase in anthropogenic CO_2 release rates over the next 50 years.
Nitrogen compounds	N_2O	N, A	S(UV)	120	300	375	350–450	Combustion and fertilizer sources.
	NH_3	N, A	T	0.01	<1	<1		Concentration variable and poorly characterized.
	$(NO + NO_2)$	N, A	T(OH)	0.001	0.05	0.05	0.05–0.1	Concentration variable and poorly characterized.
Sulfur compounds	CSO	N, A	T(O, OH)?	1(?)	0.52	0.52		Sources and sinks largely unknown.
	CS_2	N, A	T	1(?)	<0.005	<0.005		Sources uncharacterized.
	SO_2	A(?)	T(OH)	0.001	0.1	0.1	0.1–0.2	Given the short lifetime, the global presence of SO_2 is unexplained.
	H_2S	N	T(OH)	0.001	<0.05	<0.05		
Fully fluorinated species	CF_4(F14)	A	I	>500	0.07	0.24	0.2–0.31	Aluminum industry a major source.
	C_2F_6(F116)	A	I	>500	0.004	0.02	0.01–0.04	Aluminum industry a major source.
	SF_6	A	I	>500	0.001	0.003	0.002–0.05	
Chlorofluoro-carbons	$CClF_3$(F13)	A	S(UV), I	400	0.007	0.06	0.04–0.1	All chlorofluorocarbons are of exclusive man-made origin. A number of regulatory actions are pending. The nature of regulations and their effectiveness would greatly affect the growth of these chemicals over the next 50 years.
	CCl_2F_2(F12)	A	S(UV)	110	0.28	1.8	0.9–3.5	
	$CHClF_2$(F22)	A	T(OH)	20	0.06	0.9	0.4–1.9	
	CCl_3F(F11)	A	S(UV)	65	0.18	1.1	0.5–2.0	
	CF_3CF_2Cl(F115)	A	S(UV)	380	0.005	0.04	0.02–0.1	
	$CClF_2CClF_2$(F114)	A	S(UV)	180	0.015	0.14	0.06–0.3	
	CCl_2FCClF_2(F113)	A	S(UV)	90	0.025	0.17	0.08–0.3	
Chlorocarbons	CH_3Cl	N(O)	T(OH)	1.5	0.6	0.6	0.6–0.7	Dominant natural chlorine carrier of oceanic origin.
	CH_2Cl_2	A	T(OH)	0.6	0.03	0.2	0.1–0.3	A popular reactive but nontoxic solvent.

Species	Source	Removal					Remarks
CHCl$_3$	A	T(OH)	0.6	0.01	0.03	0.02–0.1	Used for manufacture of F22; many secondary sources also exist.
CCl$_4$	A	S(UV)	25–50	0.13	0.3	0.2–0.4	Used in manufacture of fluorocarbons; many other applications as well.
CH$_2$ClCH$_2$Cl	A	T(OH)	0.4	0.03	0.1	0.06–0.3	A major chemical intermediate (global production = 10 Tg/yr); possibly toxic.
CH$_3$CCl$_3$	A	T(OH)	8.0	0.14	1.5	0.7–3.7	Nontoxic, largely uncontrolled degreasing solvent.
C$_2$HCl$_3$	A	T(OH)	0.02	0.005	0.01	0.005–0.02	Possibly toxic, declining markets because of substitution to CH$_3$CCl$_3$.
C$_2$Cl$_4$	A	T(OH)	0.5	0.3	0.07	0.03–0.2	Possibly toxic, moderate growth owing to substitution to CH$_3$CCl$_3$.
Brominated and iodated species							
CH$_3$Br	N	T(OH)	1.7	0.01	0.01	0.01–0.02	Major natural bromine carrier.
CBrF$_3$(F13B1)	A	S(UV)	110	0.001	0.005	0.003–0.01	Fire extinguisher.
CH$_2$BrCH$_2$Br	A	T(OH)	0.4	0.002	0.002	0.001–0.01	Major gasoline additive for lead scavenging; also a fumigant.
CH$_2$I	N	T(UV)	0.02	0.002	0.002		Exclusively of oceanic origin.
Hydrocarbons, CO, and H$_2$							
CH$_4$	N	T(OH)	5–10	1650	2340	1850–3300	A trend showing increase over the last 2 years has been identified.
C$_2$H$_6$	N	T(OH)	0.3	0.8	0.8	0.8–1.2	Predominantly of auto exhaust origin.
C$_2$H$_2$	A	T(OH)	0.3	0.06	0.1	0.06–0.16	No trend has been identified to date.
C$_3$H$_8$	N	T(OH)	0.03	0.05	0.05	0.05–0.1	No trend has been identified to date.
CO	N, A	T(OH)	0.3	90	115	90–160	No trend has been identified to date.
H$_2$	N, A	T(SL, OH)	2	560	760	560–1140	
Ozone (Tropospheric)							
O$_3$	N	T(UV, SL, O)	0.1–0.3	F(Z)	12.5%		A small trend appears to exist, but data are insufficient.
Aldehydes							
HCHO	N	T(OH, UV)	0.001	0.2	0.2	0.2	Secondary products of hydrocarbon oxidation. 1980 concentration estimated from theory.
CH$_3$CHO	N	T(OH, UV)	0.001	0.02	0.02	0.02	

*N: natural; A: anthropogenic; O: oceanic; S: stratospheric; UV: ultraviolet photolysis; T: tropospheric; OH: hydroxyl radical removal; I: ionospheric and extreme UV and electron capture removal; and SL: soil sink.

†These concentrations are integrated averages; for chemicals with lifetimes of 10 years or less, significant latitudinal gradients can be expected in the troposphere; for chemicals with extremely short lifetimes (0.001–0.3 years), vertical gradients may also be encountered.

‡These values are not used in the present assessment.

§Also see text for details.

(*Source: Ramanathan et al., 1985.*)

FIGURE 16.3

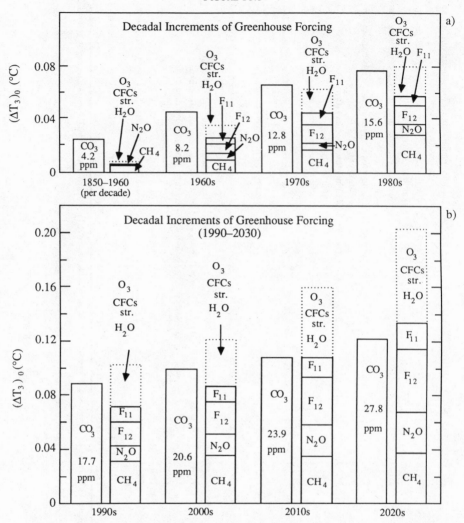

Decadal additions (see Editor's Note below) to global mean greenhouse forcing of the climate system. The $(\Delta T_s)_0$ is the computed temperature change at equilibrium $(t \to \infty)$ for the estimated decadal increases in trace gas abundances, with no climate feedbacks included. Formulas for (ΔT_s) as a function of the trace gas abundances are given by Lacis et al., 1981. (a) Past additions. Except for O_3 and stratospheric H_2O, the estimated trace gas increases are based on measurements, as discussed in the text. (b) Future additions.

Editor's Note: *This Figure depicts the greenhouse forcing (in degrees C) due to the additions of greenhouse gases per decade. The expected increases in average global surface temperatures are greater than the climate forcing because of feedbacks, primarily increases in atmospheric water vapor associated with warming. As Ramanathan notes in his text, the 1980s greenhouse forcing of about 0.08°C from CO_2, and an equal amount from the other greenhouse gases, leads to an increase in average global temperature of between 0.2 and 0.5°C per decade.*

- At the current observed rate of increase in the trace gases, the greenhouse warming of the globe increases by 0.2 to 0.5°C per decade. Hence, if emission of pollutants by human activities continues unabated, the predicted warming can exceed several degrees °C by the middle of the next century.
- The earth as a whole will be more humid and wetter.
- The warming will be amplified significantly in the polar regions due to melting of sea ice and snow cover.
- Because of feedbacks involving the radiation, water cycle, clouds, atmospheric winds, sea ice, and oceans, shifts in climate patterns can be anticipated.

OUT-OF-EQUILIBRIUM CLIMATE AND TIMING OF THE WARMING

The most important effect of the increase in the trapping of the IR radiation energy by the trace gases is to drive the climate system out of equilibrium with the incoming solar energy. Consequently, the world oceans, the distribution of sea ice, the clouds, the biosphere, and the land are all driven out of equilibrium with respect to the energy fluxes maintaining their present state.

The climate system cannot restore the equilibrium instantaneously, and hence the surface warming and other changes will lag behind the trace gas increase. Current models indicate that this lag will range between several decades to a century. However, analyses of temperature records of the last 100 years as well as proxy records of the paleoclimate changes indicate that climate changes can also occur abruptly instead of a gradual return to equilibrium as estimated by models. The timing of the warming is one of the most uncertain aspects of the theory.

CLIMATE EXTREMES

Surface warming as large as that predicted by the models for the next century would be unprecedented because the present climate is just coming out of the peak of an interglacial. The earth's climate oscillates between glacial (ice age) and interglacial periods with a temperature variation of about 5°C. The last great ice age peaked between 14,000 to 22,000 years ago and warmed to an interglacial with the peak occurring around 5,000 years ago. Hence, a human-induced global warming of

5°C during an interglacial period would be beyond the range of extreme climates that have occurred during, at least, the last million years.

MAJOR UNCERTAINTIES IN THE PREDICTIONS

The timing of the warming is one of the major uncertainties. Next is the role of climate feedbacks that govern the magnitude of the warming. One of the least understood is how clouds will respond to the climate change and how these changes, in turn, will influence the climate. The cloud feedback question arises because clouds are very powerful regulators of the radiative heating of the planet. The next potentially important feedback is the feedback between vegetation and climate. This feedback arises because the greenhouse effect significantly perturbs the cycling of water between the oceans, the atmosphere, and the continents.

CONNECTIONS WITH THE OZONE PROBLEM

Roughly 85% of the ozone resides between 12 and 50 km, the region referred to as the stratosphere. The stratospheric ozone absorbs UV radiation from the sun, and hence regulates solar energy reaching the ground. Thus a decrease .in stratospheric ozone allows more solar energy to penetrate to the ground, which would warm the surface. At the same time, ozone is a greenhouse gas, and hence a decrease in ozone would decrease the IR warming. The net climate effect of ozone changes in the stratosphere will depend on where in the atmosphere the ozone is destroyed. In addition, a decrease in stratospheric ozone would cause a severe cooling of the stratosphere and alter the stratospheric winds. The potentially strong destabilizing effect of a severe stratospheric cooling, occurring in conjunction with a lower atmosphere warming by the greenhouse gases, has not been factored into any current model studies.

Furthermore, tropospheric ozone is an important greenhouse gas. Changes in methane, carbon monoxide, and nitrogen oxides, through chemical reactions, can lead to an increase in tropospheric ozone which can cause an additional warming. The above examples are some of the emerging issues that increasingly suggest a strong coupling between the greenhouse and the ozone problems.

References for Figures

B. Bolin, B. R. Döös, J. Jaeger, and R. A. Warrick (eds.), *The Greenhouse Effect, Climate Change and Ecosystems: A Synthesis of the Present Knowledge.* Wiley, Chichester, In press.

R. A. Rasmussen and M.A.K. Khalil, 1986, *Science,* 232, 1623–1624.

A. Lacis et al., 1981, *Geophys. Res. Lett.,* 8, 1035–1038.

V. Ramanathan et al., 1985, *Geophys. Res.,* 90, 5547–5566.

Chapter 17

METHANE, CFCs, AND OTHER GREENHOUSE GASES

Donald R. Blake

Our planet is continuously bathed in solar radiation. Although we who are confined to a fixed location on the globe experience day and night, the earth does not. It is always day in the sense that the sun is shining on half of the globe. Much of the incoming solar radiation, about 30%, is scattered back to space by clouds, atmospheric gases and particles, and objects on the earth. The remaining 70% is, therefore, absorbed mostly at the earth's surface. This absorbed radiation gives up its energy to whatever absorbed it, thereby causing its temperature to increase. Be-

Testimony given in hearings before the Committee on Energy and Natural Resources, U.S. Senate, One-hundredth Congress, first session, on the Greenhouse Effect and Global Climate Change, 9–10 November 1987.

The original publication of this paper provided an illustration that has not been included here. This illustration depicts the monthly means and variances of the trace gases trichlorofluoromethane (CFC-11, or $CFCl_5$), dichlorodifluoromethane (CFC-12, CF_2Cl_2), methyl chloroform (CH_3CCl_3), carbon tetrachloride (CCl_4), and nitrous oxide (N_2O) as measured at the Barbados station of the Global Atmospheric Gases Experiment.

cause solar radiation is absorbed continuously by the earth, it might be supposed that its temperature should continue to increase. It does not, of course, because the earth also emits radiation, the spectral distribution of which is quite different from that of the incoming solar radiation. The higher the earth's temperature, the more infrared radiation it emits. At a sufficiently high temperature, the total rate of emission of infrared radiation equals the rate of absorption of solar radiation. Radiative equilibrium has been achieved, although it is a dynamic equilibrium: absorption and emission go on continuously at equal rates. The temperature at which this occurs is called the radiative equilibrium temperature of the earth. This is an average temperature, not the temperature at any one location or at any one time. It is merely the temperature that the earth, as a blackbody, must have in order to emit as much radiant energy as the earth absorbs solar energy.

All objects emit radiation continuously, mostly in the infrared region. Some of this radiation is absorbed by the atmosphere, which makes it warmer than it would otherwise be. But the warmer the atmosphere the more it emits to the ground, hence the ground is warmer than it would otherwise be. The extent to which the atmosphere absorbs infrared radiation is, therefore, of great importance in determining the temperature of the earth's surface. If the atmosphere absorbs no infrared radiation, the average surface temperature would be well below freezing, about 255 K (-18°C), and so would the air in contact with it. However, the atmosphere does absorb infrared radiation, and as a consequence surface temperatures are higher than they would otherwise be. This warming of the atmosphere and earth's surface has come to be known as the greenhouse effect.

Not all atmospheric gases absorb infrared radiation to the same degree. Indeed, the most abundant ones by far, nitrogen and oxygen, are the least absorbing. Of far greater importance are water vapor and much less abundant gases such as carbon dioxide (CO_2) and ozone (O_3). The atmospheric concentration of CO_2 is about 350 parts per million by volume (ppmv) and is increasing at a yearly rate of about 1 ppmv. [A plot of the CO_2 concentration versus time is given in Figure 5.1 in Chapter 5.] Fossil fuel combustion is the major source of carbon dioxide. About 60% of the carbon dioxide remains in the atmosphere while the rest is absorbed by the oceans or biosphere. Of major concern is whether the oceans and biosphere can maintain pace with carbon dioxide production, thus keeping the fraction at 60%. If not, the yearly increase in carbon dioxide will continue to rise.

One's first reaction about greenhouse contributions is that something like CCl_2F_2 (CFC-12), at levels of about 0.00045 ppmv and increasing annually at 0.00002 ppmv, would be trivial. However, as was previously mentioned, the greenhouse effect occurs through the absorption of infrared radiation into specific vibrational frequencies of polyatomic molecules in the atmosphere. Infrared radiation of the wavelengths which can be absorbed by CO_2 vibrations are already exposed to absorption by 350 ppmv of CO_2, and absorption by these vibrations is nearly saturated. Increasing the concentration to 351 ppmv will increase the probability of absorption a little, but it is hard to increase the total absorption very much if it is already 99.9%. On the other hand, the amounts of CFC-12, CFC-11 (CCl_3F), CFC-113 (CCl_2FCClF_2), methyl chloroform (CH_3CCl_3), carbon tetrachloride (CCl_4), nitrous oxide (N_2O), and methane (CH_4) in the atmosphere are much smaller, and absorption in their vibrational frequencies is very far from saturation. Some of these absorptions happen to fall into infrared "windows" in which absorption by H_2O, CO_2, and O_3 are all very weak or nonexistent, and the individual molecules can be as much as 10,000 times more effective in retaining infrared radiation than is the added CO_2 when the latter concentration is already about 350 ppmv. When you consider the effects from increasing concentrations of these halocarbons, N_2O, CH_4, and tropospheric O_3, etc., the combined absorptions from increments in these gases over the past decade become approximately as important as the increment in CO_2 in contributions to the anticipated overall greenhouse effect.

INCREASING CONCENTRATIONS OF SELECTED TRACE GASES

Trace gas concentrations in the atmosphere reflect, in part, the overall metabolism of the biosphere, and the broad range of human activities such as agriculture, production of industrial chemicals, and combustion of fossil fuels and biomass. There is dramatic evidence that the composition of the atmosphere is now changing, due to increased gaseous emissions associated with human activities.

Detailed measurements of various halocarbons have been conducted since 1978 by a number of research groups throughout the United States. Although the absolute concentrations of individual species may vary by as much as 10% between research groups, there is generally good agreement concerning the overall trends.

Chlorofluorocarbons 12 and 11 are the most widely used CFCs at this time. They are globally used in many and varying applications, some of which are as aerosol propellants, refrigerants, and foaming agents for plastics. Both CFC-12 and CFC-11 are increasing at about 5% per year and are now found in the atmosphere at concentrations of about 450 parts per trillion by volume (pptv) and 250 pptv, respectively.

Carbon tetrachloride is used predominantly as a chemical intermediate in the production of CFCs 11 and 12, leading to relatively little emission of carbon tetrachloride to the atmosphere. A small fraction is still used as a solvent in chemical and pharmaceutical production processes. Use as a grain fumigant is declining. The observed rate of increase is about 2%, while the current atmospheric concentration is about 130 pptv.

Stabilized methyl chloroform has been marketed since the early 1960s. Its principal use has always been the industrial degreasing of metallic or metaloplastic pieces. It is widely used for cold cleaning processes in the engineering industry. It is also used as a solvent in adhesives, varnishes, and paints where low flammability and low toxicity are important. Sales of methyl chloroform grew rapidly in the 1960s and early 1970s, when it replaced tri- and perchloroethylene and carbon tetrachloride in many industrial applications. The atmospheric concentration of methyl chloroform in 1987 is about 150 pptv and is increasing at about 7 pptv per year. Global emissions for methyl chloroform are actually higher than that of CFC-12 yet the overall rate of increase is less. This reflects the difference in atmospheric lifetime between these two species, 110 years for CFC-12 and about 7 years for methyl chloroform. The major removal process for methyl chloroform is by hydroxyl radical attack in the troposphere. The only known removal process for CFC-12 is photolysis in the upper stratosphere. The reason methyl chloroform has a relatively short lifetime is due to its carbon–hydrogen bonds. Thus, if the currently used CFCs are replaced by compounds which contain a hydrogen bond, the atmospheric lifetimes will be greatly reduced.

The most recent addition to CFC monitoring among many research groups is that of CFC-113. This compound is used largely as a solvent to clean and deflux sophisticated electronic assemblies and components. Its future use is vulnerable to competing systems, changes in electronics technology, and possible requirements to reclaim the solvent. One can see from the concentration versus time plot given in Figure 17.1 that the large-scale use of this particular species began about the time regulation of CFCs 11 and 12 began in this country. Although this is a difficult species to accurately measure, the atmospheric concentration is about 60 pptv and has a yearly increase of about 7 pptv.

Like the CFCs, nitrous oxide has a long lifetime, about 150 years,

FIGURE 17.1
Atmospheric Concentration of CFC-113: 1984–1987

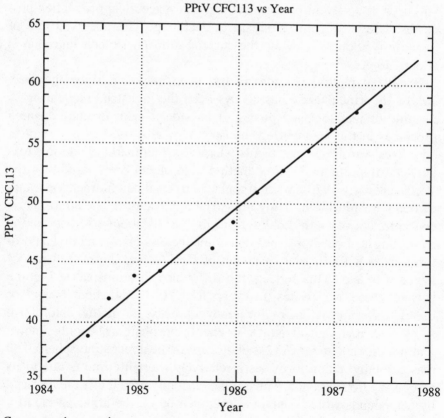

Concentration vs. time.

and is destroyed in the stratosphere. There nitrous oxide reacts with oxygen atoms to form nitric oxide, which can catalyze destruction of ozone. However, unlike the halocarbons, whose origins are unquestionably industrial, nitrous oxide has mostly natural sources. Humans may be increasing nitrous oxide emissions from soils by using ammonia and urea fertilizers. The added nitrogen eventually is returned to the atmosphere, partly as nitrous oxide. Burning coal, which contains organic nitrogen compounds, is another anthropogenic source of nitrous oxide. The rate of increase observed for nitrous oxide is about 0.3% or 1 part per billion by volume (ppbv). Currently, atmospheric levels of nitrous oxide are at about 310 ppbv.

The final greenhouse gas to be discussed here is methane. Bacteria produce methane by anaerobic fermentation in wet locations where oxygen is scarce: swamps, peat bogs, other natural wetlands, paddies, and the intestinal tracts of cattle, sheep, and termites. Oil and natural gas exploitation also may be a significant source. Studies of the carbon-14 content of atmospheric methane indicate that at least 80% must have biological origin.

Increases in atmospheric methane could be due to the greater amount of area devoted to rice cultivation or the increased number of cattle. However, increases in atmospheric methane may also be caused by rising levels of carbon monoxide from combustion processes. Both methane and carbon monoxide are removed from the atmosphere largely by reaction with hydroxyl radicals. With both gases competing for a limited amount of hydroxyl, the average lifetime of a methane molecule before it is destroyed may be getting longer. A plot of the methane concentration versus time is given in Figure 17.2. The current world average concentration of atmospheric methane is 1.685 ppmv, while the steady yearly increase observed is about 0.016 ppmv, or about 1%.

The major removal of atmospheric methane occurs in the troposphere; however, about 10% drifts up into the stratosphere. Once there, the methane is oxidized during which water molecules are formed. Two important routes exist for depositing water vapor into the stratosphere. The first is simply being carried upward as water rising in air; the second involves carrying hydrogen atoms upward in the chemical form of methane, and then releasing them as the methane molecule is oxidized. In the 1980s atmosphere, these two processes deliver

FIGURE 17.2
Atmospheric Concentration of Methane: 1978–1987

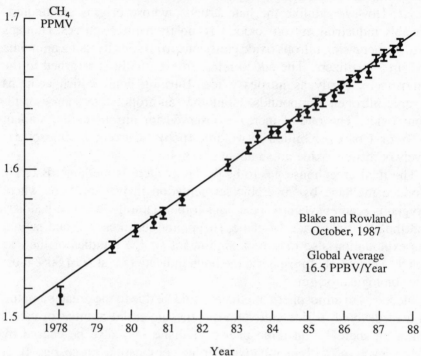

Year

Concentration vs. time. Growth in tropospheric mixing ratio of methane.

approximately equal amounts of hydrogen to the stratosphere. However, with atmospheric methane concentrations increasing at a rate of about 1% per year, the average amount of water vapor in the stratosphere is presumably steadily increasing with time. An increase in stratospheric water vapor has a binary effect. Water itself is a greenhouse gas and, thus, increasing stratospheric water vapor levels will add to the overall global warming. Secondly, the more water vapor in the stratosphere, the more water there is available to form polar stratospheric clouds, which have been linked to the dramatic losses in ozone over Antarctica during austral spring.

All of the CFCs are not only very efficient at absorbing terrestrial infrared radiation but also play a major role in transporting chlorine atoms into the stratosphere where they can participate in the catalytic destruction of the ozone layer. All of these gases have very long atmospheric lifetimes, about 110 years for CFCs 12 and 113, 75 years for CFC-11, and about 50 years for the chlorocarbon CCl_4. What this means is that these species drift around until they reach the upper stratosphere. During this lengthy journey, these compounds have been absorbing outgoing radiation being emitted by the earth and redirecting a large portion back towards the ground. Once these gases are high enough in the stratosphere they undergo another absorption, but this time it is with solar ultraviolet radiation, during which a chlorine atom is ejected from the remaining radical. At this point in time we switch from a greenhouse problem to an ozone layer problem. However, because of the extremely long atmospheric lifetimes for these halocarbons, they will have contributed to the greenhouse warming for decades or even centuries before their chlorine begins the catalytic destruction of the ozone layer.

CONCLUSIONS

Carbon dioxide, halocarbons, nitrous oxide, methane, and many more trace gases not mentioned here are all greenhouse gases that can lead to global warming. This warming is not just a theory but reality waiting to happen. These gases also play a major role in determining stratospheric ozone levels. With the unprecedented decrease in ozone above Antarctica in austral spring and recent statistical ozone studies indicating a decrease in Northern Hemisphere ozone levels by as much as 3%, it is not just a theory anymore. The gases that affect these two phenomena are rapidly increasing in atmospheric concentrations due, in part, to the extremely long lifetime of these species. At this point in time, there is no indication from the data being collected that the rate of increase for these gases will decline significantly before the end of the century. The already mentioned long lifetime of some of these gases means that even if all emissions to the atmosphere were completely halted today, many of these species would still be having an effect hundreds of years from now, on global warming and the catalytic destruction of the ozone layer.

PART V

POLICY RESPONSES

Chapter 18

THE DANGERS FROM CLIMATE WARMING: A PUBLIC AWAKENING

Rafe Pomerance

The "greenhouse effect"—this term is part of the public vocabulary now. In just a few years, it has changed from a scientific curiosity to a major policy issue for industries and governments all over the world. Why? How did a question of seemingly academic interest suddenly become the subject not only of headlines and talk shows, but of government hearings and international negotiations?

Simply put, the greenhouse effect is the process in which heat radiating from the earth's surface is trapped by gases, such as carbon dioxide and methane, in the atmosphere. The increased heat results in a rise in global temperatures which may significantly alter climate patterns. Scientists have known and studied this effect for decades, but only recently have they reached the fundamental consensus that rising levels of greenhouse gases may threaten the future of our planet. Now the implications of that possibility are reaching governments.

THE DEBATE BEGINS

The greenhouse effect reached a new stage in its evolution as a policy issue in 1979, when four eminent scientists reported to the Council on Environmental Quality (CEQ) that "man is setting in motion a series of events that seem certain to cause a significant warming of world climates unless mitigating steps are taken immediately." The authors were ecologist George Woodwell, one of the first to examine the role of deforestation in the buildup of carbon dioxide; geophysicist Gordon MacDonald, one of CEQ's original members; David Keeling of the Scripps Institute of Oceanography, who coordinated continuous measurements of carbon dioxide in the atmosphere; and oceanographer Roger Revelle, who established the carbon dioxide monitoring station at Mauna Loa in Hawaii in 1957 and who chaired the 1977 National Research Council report "Energy and Climate."

At about the same time, the National Academy of Sciences initiated a study of the greenhouse effect at the suggestion of President Carter's science advisor Frank Press. After reviewing available atmospheric models and analyses of past climates, the study panel chaired by meteorologist Jule Charney concluded that "we have tried but have been unable to find any overlooked or underestimated physical effects that could reduce the estimated global warming due to a doubling of CO_2 (carbon dioxide) to negligible proportions or reverse them altogether." The study estimated that a doubling of CO_2 in the atmosphere would raise global temperature by 3°C plus or minus 1.5°C.

The greenhouse problem was debated in yet another forum that year when the Carter administration proposed a major synthetic fuels initiative. In an editorial in The Washington Post, Gordon MacDonald argued that synthetic fuels produced even more carbon dioxide per unit of energy than coal, oil, or natural gas. MacDonald warned that subsidizing synthetic fuels was a mistake that would only increase U.S. dependence on CO_2-intensive energy systems.

The controversy attracted the interest of then U.S. Senator Abraham Ribicoff, who had recently been warned of the greenhouse effect by West German Chancellor Helmut Schmidt. Ribicoff convened a Senate symposium on the subject. The result was an amendment to the synthetic fuels legislation of 1980 mandating that the National Academy of Sciences undertake a comprehensive review of the problem. Also in 1980, the National Commission on Air Quality established by the Clean Air Act amendments of 1977 held a workshop on the greenhouse effect as part of its review of the Clean Air Act. That workshop may have been the first effort to concentrate solely on public policy issues rather than scientific aspects of the problem.

In January 1981, under the leadership of Gus Speth, the Council on Environmental Quality released an internal report on the CO_2 problem. After analyzing the reductions in CO_2 emissions that would be needed to keep levels below 1.5 times the preindustrial levels, CEQ concluded that "the potential risks from even moderate increases in the burning of fossil fuels . . . underscores the vital need to incorporate the CO_2 issue into the development of United States and global energy policy." Confirming an added dimension to the problem, scientists at the Goddard Institute of Space Studies concluded later that same year that CO_2 was not the only problem gas—methane, tropospheric ozone, nitrous oxides, and chlorofluorocarbons (CFCs) could also contribute significantly to warming the atmosphere.

The Environmental Protection Agency made its first contribution to the debate in 1983, when it released its report "Can We Delay a Greenhouse Warming?" EPA's report concluded that levels of atmospheric greenhouse gases were already high enough to trigger a global warming, and economic momentum would ensure even further warming. The report further concluded that global temperatures would rise by 2°C within a relatively short time, even with major reductions of CO_2 emissions, although such reductions could have an impact in the long run.

EPA's report was followed shortly by *Changing Climate*, the greenhouse study of the National Academy of Sciences that had been triggered by Ribicoff. In contrast to EPA's conclusions about fossil fuel use and CO_2 buildup, the academy judged that "we do not believe that the evidence at hand about CO_2-induced climate change would support steps to change current fuel use patterns away from fossil fuels." Perhaps the academy's report calmed public fears. At any rate, the issue faded from the public eye until 1985, when new scientific information, an international conference, and a series of congressional hearings combined to return the greenhouse effect to public awareness.

A NEW BURST OF INTEREST

Early in 1985, scientists V. Ramanathan and Ralph Cicerone and their colleagues from the National Center for Atmospheric Research announced that not only were other greenhouse gases contributing as much to global warming in the 1980s as CO_2, but these gases could eventually surpass carbon dioxide in their contribution to the greenhouse effect. These findings reinforced the growing consensus that some global warming was inevitable and that it would occur rapidly.

An international meeting in October 1985 came to the same conclusion. Under the auspices of the United Nations Environment Program (UNEP) and the World Meteorological Organization (WMO), scientists from 29 nations met in Villach, Austria, and agreed that "some warming of climate now appears inevitable; the rate of future warming could be profoundly affected by government policies on energy conservation, on use of fossil fuels, and emission of some greenhouse gases."

Following on the heels of the Villach Conference was a Senate hearing convened by Senator David Durenberger (R–MN), as well as a call by Senator Albert Gore (D–TN) for an international "Year of the Greenhouse" to focus global attention on the problem. Gore was not new to the issue, having conducted hearings on the greenhouse effect in 1982 and 1984 while he was a member of the U.S. House of Representatives. The pace quickened in 1986, when WMO, the National Aeronautics and Space Administration (NASA), and numerous other agencies issued a three-volume report on atmospheric ozone. The report detailed the rapid atmospheric changes occurring as a result of human activity, with particular focus on the depletion of the protective ozone layer in the stratosphere and on the greenhouse effect. The report concluded: "There is now compelling evidence that the atmosphere is changing on a global scale." Finally, hearings held by Senator John Chafee (R–RI) in June 1986 brought together key scientists and government officials to discuss the problem. Perhaps the most significant testimony came from Dr. James Hansen of the Goddard Institute for Space Studies. Based on his climate models, Hansen projected that significant warming might be observed within 5 to 15 years.

This was a surprise to many observers. The greenhouse problem had been viewed as taking decades to develop, and, indeed, doubled levels of carbon dioxide in the atmosphere were still projected to occur decades later. That warming could occur at much lower levels of CO_2 had rarely been suggested by the scientific community. Now, suddenly, global warming had become a serious issue for government policy-makers.

The Chafee hearings transformed the priority of the greenhouse issue, making it more important in policy discussions. Chafee moved the issue another crucial step by asking EPA to develop a set of policy options for stabilizing the level of greenhouse gases in the atmosphere. For the first time, members of Congress were calling on the executive branch to complete an analysis of what governments might do about cutting greenhouse gas emissions.

Moreover, this was the first congressional forum at which the greenhouse issue and ozone depletion were examined at the same time. The committee heard that both issues were planetary in scale. Both problems concerned the fate and well-being of fundamental life support systems of the earth. And the solutions were coupled because CFCs are both a greenhouse gas and an ozone-depleting chemical. The joining of two issues with such profound implications caused a dramatic increase in the media's attention. An immediate effect of the hearings was apparent within weeks when Congress made a substantial addition to the appropriations for EPA's policy work on climate change and ozone depletion.

The Chafee hearings also provided political support for a critical decision made later in the year. In early November, EPA Administrator Lee Thomas announced that the United States government would support a phase-out of CFCs as the international negotiations got under way on the future of the ozone layer. After a few months of negotiating sessions, 24 nations agreed to a 50% cut in CFC consumption in industrialized nations by 1999. This agreement, known as the Montreal Protocol on Substances That Deplete the Ozone Layer, was signed in September 1987. The accord had immediate and obvious implications for the greenhouse effect since CFCs make up 15 to 20% of current additional greenhouse warming. To many observers, the Montreal Protocol also became a model of what an agreement to control carbon dioxide and other greenhouse gases might look like.

The Senate took up the greenhouse effect and ozone depletion issues again in January 1987. In fact, it became the subject for the first major hearing by the Senate Environment and Public Works Committee in the new Congress. Two critical and relatively new problems were discussed at this hearing that were to become central aspects of the growing urgency associated with the global warming problem.

Ramanathan argued in the hearing that atmospheric greenhouse gas concentrations had already been altered sufficiently by 1980 to commit the earth to a 0.7 to 2°C warming. With each passing decade Ramanathan estimated that an additional 0.2 to 0.5°C was being added. His analysis meant that by the year 2020—in 33 years—the earth would be committed to as much as 4°C warming. Many scientists believe that the earth has not been 4°C warmer for tens of millions of years. Ramanathan's testimony established that society was already locked into a substantial amount of climate change no matter what governments did. The problem was no longer a question of whether a change would occur but how much and when.

The second major issue was raised by Wally Broecker, a geochemist at the Lamont Dougherty Laboratory. Broecker's testimony was a follow-up to a talk he had given at an EPA conference in June. Broecker said that an examination of the history of climate change suggested that the greenhouse effect might push the earth into a state of rapid change—reorganizing the earth systems in the process. Broecker had little faith that society would experience a linear and gradual change in global temperature and climate as suggested by general circulation models of the atmosphere. The key implication of Broecker's testimony was that the buildup of greenhouse gases could force the climate system to go into a state of rapid change and that society ultimately had limited ability to predict what that change might bring.

In 1987, attention turned to another forum in Congress when the Senate Energy Committee took up the issue. This effort began in January when the chairman of the Energy Committee, Senator Bennett Johnston (D–LA), received an expert's briefing. As a result, a two-day hearing, chaired by Senator Tim Wirth (D-CO), was held in November. Two policy suggestions made at the hearing took hold. Gordon Mac-Donald suggested that consideration be given to a carbon tax to discourage use of carbon-based fuels and encourage more efficient use of energy. Some months later, this proposal resulted in a letter from Johnston and Wirth to the Congressional Budget Office (CBO) requesting that CBO undertake a study of the feasibility of a carbon tax. The other was a suggestion by Gus Speth, president of the World Resources

Institute, that Congress ought to consider comprehensive legislation on global warming. Wirth took up the challenge and 10 months later introduced comprehensive legislation on the global warming problem.

At about the same time there was a legislative surprise in the Senate when an amendment was attached to the State Department Authorization Act in the Senate Foreign Relations Committee by Senator Joe Biden (D–DE), who had taken an interest in the issue during his presidential campaign. Biden's amendment was amended on the Senate floor due mainly to the work of Senators Max Baucus and John Chafee. After substantial negotiations a somewhat altered version of Biden's amendment survived a House–Senate conference and was signed into law by President Reagan in January 1988. The amendment, known as the Global Climate Protection Act, was designed to require the executive branch to deal with the "prevention" aspects of the problem. Its most significant provisions obliged the president to submit to Congress a plan for stabilizing the concentration of greenhouse gases in the atmosphere. The legislation established global warming as a greater priority in U.S. foreign policy, particularly with respect to U.S./Soviet relations.

PROGRESS ON THE INTERNATIONAL FRONT

Nineteen eighty-seven was also a big year for the greenhouse effect internationally. Since 1972, the United States and the Soviet Union had carried out a successful bilateral agreement to study a number of environmental issues, including climate change. As a result of suggestions made by Senator Chafee and Gus Speth, negotiations were begun among Lee Thomas, administrator of the EPA, his counterpart in the Soviet Union, Yuri Izrael, and Alan Hecht, head of the National Climate Program Office (NCPO) of the National Oceanic and Atmospheric Administration (NOAA) to establish a new U.S./Soviet working group to look at possible "response strategies" to the greenhouse effect. An agreement was reached in early 1988 to begin that effort.

At his June 1986 hearing and in follow-up letters, Chafee and other members of the Senate had urged that President Reagan raise the issue at the next U.S./Soviet summit. An important paragraph urging continued cooperation on the issue was agreed to in the joint Reagan/Gorbachev communiqué issued at the conclusion of the Washington Summit in December 1987. Both sides agreed to continue their joint studies.

In June 1987, the UNEP Governing Council met in Nairobi, Kenya. A number of delegations were pressing UNEP for more action and a commitment to negotiate a greenhouse convention. Such proposals were resisted as "premature" by the United States and a number of other nations. The Governing Council agreed that the executive director should report back to the council in 1989 on governmental policy options to deal with the greenhouse effect. In a joint effort with the World Meteorological Organization, the UNEP Governing Council also decided to establish an intergovernmental scientific assessment of the global warming problem. The first meeting of the joint UNEP/WMO intergovernmental body eventually took place in November 1988 in Geneva, some 16 months after it was agreed to.

Events were moving in West Germany as well. In response to a citizens' meeting held in late 1986—just prior to an upcoming national election—the West German government decided to take a leading role in the ozone negotiations and established in early 1987 a special committee of the West German Parliament of parliamentarians and scientists to examine the greenhouse effect and ozone depletion problem.

Initiatives were also taken within the British Commonwealth, led by the tiny island nation of the Maldives which would be threatened with destruction if a sea-level rise of a few feet were to occur. Consequently, a special study group on the greenhouse effect was established. This step eventually helped to stimulate a key speech given by Margaret Thatcher to the Royal Society in October 1988 in which Thatcher, for the first time, acknowledged the importance of the global warming problem.

A more policy-oriented international assessment was coordinated by the Beijer Institute of Stockholm for UNEP. Meetings were held in Villach, Austria, and Bellagio, Italy, to formulate an up-to-date assessment of the climate change problem. The UNEP/WMO report "Developing Policies for Responding to Climatic Change," released in April 1988, was also designed to set the stage for the Toronto Conference, "The Changing Atmosphere: Implications for Global Security," in June 1988. The UNEP/WMO report's major conclusion was that global warming would occur at a more rapid pace than climate change of the past. Many natural systems simply would not be able to survive such a rapid alteration of climate conditions. The study therefore urged that the rate of climate change be slowed to a pace to which natural systems could adjust.

As the UNEP/WMO report was being completed in the spring of 1988, the news was again dominated by the ozone layer. A scientific panel organized by NASA and NOAA concluded that there had been a loss of ozone over high latitudes in the Northern Hemisphere—a loss that could not be attributed to natural variability. The largest loss was apparent at polar latitudes, which suggested the possibility of an Arctic ozone hole. The report also concluded, as a result of two missions to Antarctica in late summer and early fall of 1986 and 1987, that the Antarctic ozone hole is caused by chlorofluorocarbons.

A few days before the Ozone Trends Panel report was released, the Montreal Protocol was ratified by the U.S. Senate. Many members of Congress and environmentalists, however, realized that the Montreal Protocol was simply not sufficient to protect the ozone layer. Indeed, the Montreal Protocol had been negotiated before we learned of the ozone losses over the Arctic and discovered that the Antarctic ozone hole had been caused by CFCs. Under the Montreal Protocol, CFC concentrations were still expected to increase the risk of further ozone depletion. As a result of this announcement, UNEP moved up the date of the first review and possible strengthening of the Montreal Protocol. That meeting will take place in April 1989.

At the same time, the initiatives started in the Senate Energy Committee were approaching fruition. Senator Wirth had taken up the challenge to develop comprehensive legislation on the climate change problem. The legislation was in the final stage of development when Wirth attended a briefing organized by the Energy and Environmental Study Institute on the recently released UNEP/WMO/Bellagio report. Wirth decided to convene a hearing on the report but at the same time decided to focus attention on the drought and heat wave which were enveloping much of the continental United States at that time. The drought, Wirth believed, dramatized the implications of climate change. The hearing was scheduled just days before the start of the Toronto Conference.

The central focus of the hearing turned out to be the testimony of Dr. James Hansen of the Goddard Institute for Space Studies, who testified that global temperatures had increased to the point where they were outside the range of natural variability. Hansen believed the chances were extremely high that such an excursion of global temperatures was due to the greenhouse effect. Hansen also testified that his general circulation model experiments showed that summer heat waves and drought were more likely in certain regions of the United States as the earth warmed. Although Hansen's testimony did not specifically link the drought and the heat wave of 1988 to the greenhouse effect, he drew significant attention to the whole issue.

The Senate hearing was followed the next week by the Toronto Conference, where heads of state addressed the issues. At the opening of the conference, the prime minister of Norway, Gro Harlem Brundtland, and the prime minister of Canada, Brian Mulroney, both appealed for a global convention to deal with the greenhouse effect. No head of state had ever made such a call before. The conference adopted a statement that called for industrialized nations to cut their CO_2 emissions by 20% by the year 2005. For the first time a group of international experts had focused on specific cuts in CO_2 emissions. Senator Wirth, the keynote speaker at the Toronto meeting, proposed a 20% cut in CO_2 emissions by the year 2000 as did a statement prepared by nongovernmental organizations at Toronto.

Within days of the conclusion of the Toronto meeting, Senator Wirth introduced his comprehensive legislation on climate change. At the same time, Senator Robert Stafford (R–VT) introduced legislation aimed at reducing CO_2 emissions. Within a few weeks another shift had occurred in the greenhouse debate—from a focus solely on CFC reduction to a major call for CO_2 reductions. In October 1988, Congresswoman Claudine Schneider introduced a third version of comprehensive legislation on the problem. After the introduction of the Wirth and Stafford bills, the summer of 1988 continued to cause an enormous amount of press coverage on the greenhouse effect with attention focused on what could be done to reduce CO_2 emissions.

A NEW LEVEL OF RECOGNITION

The issue even reached the presidential campaign. During the primaries, the Democratic candidates debated environmental issues in New Hampshire. At that debate the future Democratic nominee, Governor Michael Dukakis, pledged to convene a global summit on the problem and to sit down with Soviet leaders on the issue. Dukakis reiterated this pledge and endorsed a global convention at a speech in Rocky Mountain National Park just before the Democratic convention.

George Bush went on a campaign swing on environmental issues shortly after the conclusion of the Republican convention. Bush made a major speech on the environment in which he pledged to convene a high-level international meeting on global environmental issues including the greenhouse effect. Bush promised to come up with "an action plan" at that conference to start solving the problem. In the vice-presidential debate, both Lloyd Bentsen and Dan Quayle were asked whether they were prepared to reduce the use of fossil fuels to deal with the greenhouse effect.

Statements of the presidential candidates gave observers a clear indication that the issue had reached a new level of recognition. Campaign speeches, Sunday morning talk shows, dozens of newspaper editorials, numerous magazine stories—all indicate that the greenhouse effect has arrived as a major policy problem. Four government studies have been initiated to examine policy options on CO_2. Three comprehensive bills have been introduced and more are on the way. Forty-two senators and two heads of state have called for a climate convention. Conferences and briefings are continually scheduled. The greenhouse effect is rapidly making its way onto the foreign policy agenda.

In 1989 the Canadian government is planning a meeting on a possible "law of the atmosphere." The Dutch environmental minister is convening a ministerial-level meeting on the greenhouse effect. The UNEP Governing Council will convene to explore policy options and possible steps toward a convention. President Bush has pledged a high-level meeting on the greenhouse effect to be held at the White House. The ozone negotiations will resume.

Since the Chafee hearings in 1986, the greenhouse effect has continued to gain interest and attention at an extremely rapid pace. Nineteen eighty-nine will be the year in which the focus turns to the real question: What are we going to do about it?

Chapter 19

ONE LAST CHANCE FOR A NATIONAL ENERGY POLICY

Ralph Cavanagh, David Goldstein, and Robert Watson

America's energy affairs are rich in irony and paradox. We could do with much less of each, but what follows is intended more as a corrective than a condemnation. The aim is to point the way toward a sound and broadly acceptable national energy policy, which builds on the successes and failures of four presidential administrations since 1973. A few events of recent vintage provide some instructive context.[1]

The abandonment of New York's five billion dollar Shoreham power plant in mid-1988 prompted extensive publicity and much editorial handwringing about potential electricity shortages, even as largely

From a statement released to the presidential candidates, July 1988, by the Natural Resources Defense Council.

overlooked national appliance efficiency standards were relieving utilities inexpensively of the need to finance and build more than 35 Shoreham-sized generators.[2] Yet a president who had vigorously supported Shoreham tried to veto the appliance legislation. In its last full budget cycle (1987–1988), his Department of Energy spent more money on nuclear power than on all other technologies for producing and saving energy combined—even though the last order for a nuclear plant had been placed in 1978, and all plants ordered since 1974 had been canceled.[3]

In January 1988, the same Department of Energy requested a 50% reduction in a national energy conservation research budget that previously had been cut almost 70% from 1980 levels,[4] while the Department of the Interior was battling environmentalists over a massive program for extracting new energy supplies from sensitive offshore and Alaskan areas. Yet the officially anticipated oil supply from all these fields totaled less than half of a first installment on unexploited oil *savings* from efficiency improvements in American homes and transportation systems.[5] And the savings would continue to grow long after the fields had stopped producing, displacing fossil fuel consumption that is the largest single contributor to the continuing "global warming" that threatens every nation with ruinous climate changes.

What kind of national energy policy produces results and misses opportunities like these? The question itself might have seemed inappropriate as recently as the early 1970s. For three decades following World War II, the U.S. government made little pretense of creating or executing national energy policy. To be sure, the federal government financed numerous highways, issued licenses for selected classes of power plants, leased mineral resources on public lands, manipulated the prices of assorted fuels, and kept the tax code hospitable to those in the business of producing oil, gas, or electricity. Several federal agencies were even in that business themselves, mostly to sell hydropower from publicly financed dams. But these functions were acquired haphazardly and scattered throughout the government, with no semblance of coordination or integrated planning.

From 1973–1980, three administrations made concerted efforts to change all that. They enacted reams of statutes, appropriated tens of billions of dollars, and marshaled small armies of scientists in a quest for a more stable, secure, and environmentally benign energy future.

Enough time has passed to permit an assessment of what their record can teach a national leadership that includes a new president.

Before 1973, history records no groundswell of protest at the federal government's inability or unwillingness to guide the nation's energy affairs. Fuels of all kinds were cheap and getting cheaper. The public paid little attention to commodities that, after all, had no value independent of the services that they provided. Aside from the occasional eccentric who enjoys the smell of gasoline, no one desires any form of energy for itself. What matters are the services that energy provides: heat, cooling, dishwashing, steel-making, night baseball. Absent substantial effects on quality or cost, the source and quantity of energy consumed to produce such services generally will be a matter of sublime indifference.

Among those most visibly indifferent to energy considerations have been the designers of America's buildings, appliances, vehicles, and industrial processes. The typical household refrigerator of the mid-1970s used five times as much electricity as its late 1940s counterpart.[6] The average car on the road in 1973 got 10% fewer miles per gallon, compared with the average for 1961.[7] Most houses built prior to the 1970s had single-pane windows and little or no insulation. Trends like these pushed energy consumption up more than three times as fast as population growth from 1949–1973.[8] Domestic resources could not keep pace with this surge, and by 1973 one-fifth of the energy consumed in the United States was imported. Half of that was oil purchased from an aspiring but ineffective cartel called the Organization of Petroleum Exporting Companies (OPEC).

The first serious concerns about rapid growth in energy consumption emerged in the 1960s from a renascent environmental community. Energy production was a leading cause of air pollution. It was the civilian economy's dominant source of radioactive waste and the impetus behind most efforts to dam free-flowing rivers. In 1965, a court battle over a proposed Hudson River hydroelectric plant at Storm King Mountain helped launch the modern environmental movement.[9] Later in the decade, debates that continue to this day were framed by a major oil spill in California and a clash over Alaska's Prudhoe Bay oil fields and pipeline.

Many opponents of oil and hydro development were at least equally concerned about coal mining and combustion, which threatened human and other environments in diverse ways. Nuclear options were suspect on public health grounds, including unresolved waste disposal issues and potential links between civilian and military applications of atomic energy. Natural gas seemed relatively benign, but both applications and safely accessible supplies were limited.

A yearning for alternatives to these energy resources became almost universal when OPEC taught Americans a sharp lesson in 1973. Following the initial spate of price hikes and fuel shortages, Congress officially discerned that "the nation is currently suffering a critical shortage of environmentally acceptable forms of energy."[10] In words that still linger on the statute books, the national legislature proclaimed that "the urgency of the Nation's energy challenge will require commitments similar to those undertaken in the Manhattan and Apollo projects."[11]

As this rhetoric suggests, the federal government initially treated the national "energy crisis" as primarily a technological problem, to be solved with new and better hardware created by experts pressed into service under "crash" federal programs. One of the first enactments directed a single luckless administrator to prepare a plan of research and development

designed to achieve—
(1) solutions to immediate and short-term (to the early 1980s) energy supply system and associated environmental problems;
(2) solutions to middle term (the early 1980s to 2000) energy supply system and associated environmental problems; and
(3) solutions to long term (beyond 2000) energy supply system and associated environmental problems.[12]

All of these "solutions" were complicated by a general consensus that economic growth was impossible without at least equivalent increases in energy use. On that view, 20 years of even modest prosperity—say, economic growth averaging around 2% per year—implied almost 50% greater energy requirements at the end of the cycle. To make matters worse, electricity was steadily increasing its share of the national energy budget, and disproportionate fractions of the energy used to produce electricity (generally at least two-thirds) were lost as waste heat to the surrounding environment. As a result, growing electricity needs brought with them disproportionate additions to the nation's total energy needs. Under all these circumstances, how could a nation already dangerously dependent on imports find energy security, environmental quality, and economic growth in any of the periods that the Congress had identified in its mandate?

To Congress itself, the answer was to encourage the development of essentially every energy production option that anyone reputable would mention favorably to a House or Senate Committee (and at one point in 1979, energy legislation was pending before 78 committees in the House alone, leaving 421 of its 435 members at least nominally responsible for abating the energy crisis).[13]

In various enactments, legislators went on record in vigorous support of solar heating and cooling, wind energy, electricity and steam heat from geothermal resources, small-scale hydropower, use of waste heat from industrial processes to make electricity ("cogeneration"), wood-fired power plants, photovoltaic cells (which convert sunlight directly into electricity), solar thermal generators (which tap steam produced by boiling water with focused rays of sunlight), fusion power (which recreates the sun's own internal energy source), tidal power, "ocean thermal energy conversion" (a process that exploits the temperature changes at different ocean depths), alcohol-based fuels, and various "synthetic fuels" (produced by extracting flammable gases and liquids from various plentiful minerals, including coal and oil-saturated rocks).

Each item on this far from complete list had fervent champions in the legislative branch and the new cabinet-level Department of Energy, which Congress created in 1977 to carry out national energy policy. Each alternative avoided reliance on foreign supplies, and many used fuels that were self-renewing. The very length of the list was comforting to many; surely in so diverse a portfolio could be found hedges against every possible calamity. Our inventory of conventional power plants, petroleum reserves, and natural gas supplies would provide the bridge into an era of plentiful, inexpensive, and renewable new energy resources. As explained in 1974 by Donald Hodel, who later became secretary of energy: "Every civilization must go through this. Those that don't make it destroy themselves. Those that do make it wind up cavorting all over the universe."[14]

Fourteen years have passed since that utterance. We haven't made it. Both the extent of our failure and its explanation are instructive, as are some remarkable successes that the Congress of the 1970s never envisioned.

Despite the teachings of some venerable and revered fables, there are times when a great many small contributions to a worthy goal add up to no more than a small contribution. So it has been with alternative energy resources. Taken together, the dozen-odd resources listed earlier met at most 3% of U.S. energy needs by the mid-1980s, and little of that could be traced to technological advances spurred with federal resources. This hardly vindicates hopes, once widely voiced within the government, that solar energy alone "could provide about 20 percent of the nation's energy by the year 2000."[15] And more than half of current alternative energy contributions reflect increased industrial cogeneration, an "innovation" that dates back to the dawn of the electric age.

Some of what appeared to be failed experiments were considerably, and instructively, more encouraging than others. When the federal government shut down three large wind generators in Washington State, after only one-fifth of a 30-year design lifetime, the taxpayers wrote off $55 million in investment—but could point to more than 16,000 improved applications of the same technology, which were becoming a familiar part of the western landscape.[16] In sharp contrast, the nation's first commercial synfuels plant cost $2 *billion* to build; when the sponsors defaulted on their obligations one year after starting production, the federal government had to come up with $1.64 billion in loan guarantees.[17] And the nation can be searched in vain for a generation of new and improved synthetic fuels plants.

Still, exercises like the wind/synfuels comparison remain studies in largely unrealized expectations for federal energy initiatives. What went wrong with so much that once appeared so promising?

In many if not most instances, nothing approaching a fair test was ever administered. The election of Ronald Reagan in 1980 signaled a wholesale federal retreat from the energy business. The measure of this withdrawal is most cogently captured in dollars: between 1981 and 1987, federal financial support for small hydropower, wind power, solar electricity, and energy from wood and geothermal sources dropped by 80%.[18] If these cuts reflected rigorously documented deficiencies in the technologies themselves, the evidence was never published.

What was left in the "alternative energy" budget wouldn't buy the booster rockets for a space shuttle, let alone match the effort to land men on the moon. The new administration's official description of its "strategies of national energy policy" began with a pledge "to minimize federal control and involvement in energy markets."[19] Seldom have presidential aspirations been realized more thoroughly. The one partial exception was nuclear power, which remained a conspicuous if dwindling presence on the federal dole even in the twilight of the Reagan years.

On one view, the "alternative resources" sputtered because they could not survive the wholesome discipline of a free market. But the energy markets of the 1980s hardly answered to that description. In all sectors, they were dominated by production from gigantic projects that had been expensive to build but were relatively cheap to operate. Once a large-scale power plant or oil platform or gas field has been built, the cost of producing an extra kilowatt-hour or barrel or therm is generally modest compared to that for a completely new facility—even if the facility could be constructed and operated more cheaply than a new conventional plant. As long as existing conventional capacity remained less than fully utilized, the owners of this capacity could undersell sponsors of unbuilt machines, whatever their quality.

When alternative technologies did get a fair chance to compete against new conventional development, however, they scored some impressive successes. A federal law passed in 1978 guaranteed independent small-scale power producers a chance to meet utilities' needs for new generators, if they could produce electricity more cheaply than the facilities that the utilities would otherwise buy. The conclusion that emerges powerfully from this law's first decade is that if independent producers are offered long-term contracts at or near the cost of a new utility-sponsored coal or nuclear plant, enough responses promptly emerge to displace all such plants from acquisition schedules. By 1986, jurisdictions as diverse as California, Idaho, and Maine had indefinitely deferred all new coal and nuclear plants by signing up 1,424 wood-fired, hydro, wind, and cogeneration units with an average capacity of about 12 megawatts.[20] By comparison, the average nuclear plant is a 1000-megawatt brontosaurus, which cannot keep pace with the newly evolved mammals that scurry about in its shadow.

Such competitions were rare in the 1980s, however, because few energy companies were seeking new capacity. Of course, under such circumstances one would not expect to see much new energy development of any kind. And indeed conventional resources have fared at least as poorly as the alternatives in the 1980s. By 1988, domestic oil and gas production was down more than 10% from 1973 levels.[21] All nuclear power plants ordered since 1974 have been canceled, and the total bill for 115 abandoned plants exceeds $20 billion.[22] Orders for coal-fired units, which had been averaging 36 per year over the first 5 years of the 1970s, dropped to 4 per year for the same span of the 1980s.[23] Cancellations of coal units outpaced orders by more than two to one over this period. We appear to have produced an energy marketplace in which nothing can succeed.

There is a conspicuous exception to this rule, which can be captured in one remarkable finding: energy consumption did not increase from 1973 to 1986, although the U.S. economy grew by more than 30% over that period.[24] If energy needs had simply kept pace with economic growth, almost 40% of our consumption would now have to be met with imports; by comparison, at the height of the 1973–1979 "energy crisis," we relied on outside sources for about one-fourth of our supplies.[25] Instead, imports for 1987 were lower in relative terms than those for 1973, and even electricity needs were growing less rapidly than the economy as a whole.[26]

By the federal government's reckoning, an uninflated dollar of U.S. economic growth required 25% less energy in 1986 than in 1973.[27] A 1988 *Scientific American* analysis pegged the resulting reductions in the national energy bill at *$150 billion per year,* which represents nearly $2,000 for every household.[28] Reduced energy intensities are now "producing" at least twelve times as much energy as all the Congress' "alternative resources" combined. In large measure, the general abundance of energy supply and weakness of energy prices reflect that achievement.

This happy result has had virtually nothing to do with the executive branch of the federal government, however. A case study from our national energy conservation success story helps make that point, along with others of at least equal importance.

The largest single electricity user in most American homes has been the refrigerator. In the immediate aftermath of World War II, the typical refrigerator had about seven cubic feet of usable space, a tiny freezer compartment, and an insatiable appetite for manual defrosting. Over the next quarter century, this rather unappealing appliance more than doubled in size, added a much larger and colder freezer unit, and acquired a universally acclaimed automatic defrosting cycle. Manufacturers also cut production costs by reducing the efficiency of installed equipment, so that even the small manual-defrost refrigerator of the 1970s used 70% more energy than its 1940s counterpart.[29]

These trends, coupled with growth in the number of households, drove U.S. refrigerators' electricity consumption up by almost 10% per year between 1946 and 1974. At that rate, quantities double every 7–8 years. Refrigerators needed the equivalent of 34 large coal-fired plants' power production in 1974; if the postwar rate of growth had persisted for another two decades, that total would have swelled to 215.[30]

California, which accounted for one-fifth of the nation's refrigerator market, found this and analogous forecasts unacceptable. California's utilities were facing the prospect of having to build a giant power plant along every 8 miles of coastline over the next 20 years, just to keep pace with predicted growth in electricity demand.[31] Part of the state's response was to adopt refrigerator efficiency standards, which phased in a requirement that models for 1980 would have to use at least 20% *less* electricity, on average, than their 1975 counterparts.

From 1978–1981, the California requirements became *de facto* national standards as the industry retooled to deliver equivalent frost-free cooling for less energy. The standard proved easy to meet with improved insulation materials and modest upgrades in fan and motor efficiencies. By the mid-1980s, electricity consumption per refrigerator had dropped more than 35%. A second generation of California standards paved the way for national legislation, passed in 1987, that will cut these needs by another 15–20% as of 1990, and regulators in both California and Florida are pressing for a further 30% savings by 1993. The net result should be national refrigeration energy needs in the year 1995 that are less than one-fifth of the level to which postwar trends were pointing in the mid-1970s. Rather than needing more than 107,000 megawatts of power plants, which exceeds the current capacity of the entire U.S. nuclear industry, we can expect to sustain a growing population's frost-free cooling for around 21,000 megawatts.

A conservative calculation of what we will spend on better refrigerators in order to avoid power plants yields almost five dollars in benefits for every dollar of costs, and a societal profit of more than $47 billion.[32] But we're not yet finished finding better ways to preserve food: frost-free refrigerators have been built that beat the most stringent proposed government standard by almost 20%, and technically feasible savings are more than twice this large.

Similar stories can be told about many other energy uses, including automobiles, appliances, buildings, and industrial processes. Volvo has built a crash-worthy car that accelerates faster than the typical new U.S. vehicle—and gets about 70 miles to a gallon in combined city and highway driving.[33] The developers of this four-passenger, multi-fuel model contend that "in mass production [it] would cost the same as today's average subcompact."[34] This "LCP 2000" is one of a family of 4–6 passenger prototypes, representing the work of at least five manufacturers, that have already recorded fuel efficiencies in the 70–100+ miles per gallon range.[35]

Back at home, invigorating showers can be delivered by the best inexpensive low-flow showerheads with less than half the water and water-heating energy that the average model needs. At the office, improved designs and technologies allow savings of 75% and more in providing attractive lighting.

Other opportunities abound for getting more service out of less energy. Often a saved kilowatt-hour or therm or barrel of oil is much cheaper than an additional unit of energy production. The last decade has seen extraordinary progress in the state of the art for energy efficiency in buildings, appliances, lights, and industrial processes. Entrepreneurs have demonstrated repeatedly that high-quality energy services can be delivered at 30 to 90% less energy consumption, compared with practices still common in the late 1980s.

Yet, as the refrigerator example itself suggests, we cannot simply sit back and wait for invisible economic forces to exhaust the national pool of inexpensive savings. It took the prodding of government standards to get refrigerators built more efficiently, and a similar story can be told for the nation's automobiles and many of its buildings. Often the prodding has taken the form of incentive payments or penalties rather than regulations, and sometimes both have been used together.

Obstacles to making inexpensive conservation happen include more than just inertia. When people and businesses can be persuaded even to consider investing in energy efficiency, they demand a much higher return on their money than energy companies earn on their energy production projects. Conservation typically has to recoup its full costs in three years or less, while a decade or more must pass before many new oil fields, coal mines, and power plants begin to earn *any* income, let alone profits.

There are many rational explanations for the conservers' relative stinginess. For example, decisions about efficiency levels often are made by people who will not be paying the ensuing energy bills. Sometimes what looks like indifference to efficiency is the result of inadequate information or time to evaluate it, as all will attest who have ever replaced a water heater, furnace, or refrigerator on extremely short notice. Whatever its sources, the gap between the investment perspectives of conservers and producers invites unnecessary expenditures on energy facilities that could have been avoided less expensively with efficiency improvements.

These barriers to improved energy efficiency create some special problems for a society whose government thinks that the free market has the answers to all energy questions.

Our achievements in energy conservation, considerable though they are, have only scratched the surface of the possible. We have dragged our fleet of automobiles up to an 18 miles per gallon average in a world where 70 miles per gallon is possible. Our refrigerators are down to 1,000 kilowatt-hours per year when they could be at 300. Most of our new commercial buildings are still wasting at least half their lighting energy consumption. The list could be extended indefinitely, and it pervades the entire economy. In relative terms, we are still twice as wasteful as the Japanese, and we fell further behind them over the past decade even as our efficiencies improved substantially; they did better.[36] At the same time, we have choked off investment in an array of potentially improved energy production techniques and lost progress toward more sustainable ways of producing electricity and fuels.

We are left with a level of energy consumption that is still dangerous, on both security and environmental grounds, and a dearth of good alternatives to conventional technologies for serving it. In a world where Persian Gulf instability, acid rain, and global warming are increasingly dominant concerns, we cannot afford just to hold energy consumption stable. Consider the following:

- More than half our electricity is made with coal, which releases substantially more carbon dioxide when burned than any other utility fuel. The only known technologies for capturing these emissions are staggeringly expensive. Carbon dioxide is the most important contributor to the gradual warming of the earth's atmosphere, and fossil fuel combustion accounts for nearly three-quarters of current emissions.[37] Projections from the National Academy of Sciences and the United Nations warn of ruinous sea-level rises within decades, coupled with unprecedented disruption of global agriculture.[38] The best available data suggest that the world can check a potentially disastrous trend with significant but clearly feasible improvements in the current pace of improvements in the efficiency of energy use.[39] As the source of one-fourth of the world's energy consumption, the United States cannot go on abdicating leadership in a campaign that ultimately must enlist every nation.

- Despite more stringent pollution controls for new generators, older power plants still account for more than two-thirds of sulfur dioxide emissions, which are a principal constituent of the "acid rain" that is disrupting ecosystems throughout North America.[40] Thousands of lakes and streams across the U.S. and Canada are lifeless as a result, losses in human health, crops, and forests are mounting, and even exterior damage to buildings and monuments is now counted in the billions of dollars annually.

- If oil use simply remains constant, stemming a recent trend of increases following the 1986 OPEC price reductions, we will exhaust our readily accessible domestic reserves before 2020, even if we continuously import 40% of our needs. Throughout that process, we will become steadily more dependent on the Persian Gulf region, which already holds more than 60% of the world's proven reserves.

- Many of the energy conservation opportunities now blocked by market barriers will be lost irretrievably if we do not find ways to exploit them promptly. For example, failures to build certain efficiencies into the design of long-lived buildings and appliances are often effectively irreversible for the lifetime of the items in question. With refrigerators and furnaces lasting for 20 years or more, and building lifetimes measured in one or more half centuries, decisions to forgo savings are made not just for us, but for our children and grandchildren as well.

The federal government will not resolve these dilemmas by deferring to grossly distorted "free markets." It cannot rely solely on state or private initiative to take on problems of national if not global scale. Nor, finally, can it afford to repeat the 1970s flirtation with every conceivable "alternative" energy technology that offers some prospect of environmental and security advantages. There is a much better way to proceed, and no better time to begin than with the arrival of a new administration and Congress in January of 1989.

The clearest lesson of the last fifteen years may be that energy supply can be expanded either by finding more fuel supplies or wasting less of the supplies we already have. Recall again the modern history of the refrigerator. For purposes of meeting the needs of a growing economy and population, a kilowatt-hour preserved from waste by an efficient appliance is indistinguishable from a kilowatt-hour delivered to customers by a new power plant. Energy savings created in large quantity on a predictable schedule are energy resources, just like generators or oil fields or gas wells. If we see a need for increased supply, we should be weighing our conservation options against our generators, oil fields, and gas wells, and picking the best buys first.

This arrestingly simple principle has never figured in national energy policy. The federal government passed with lightning speed from pursuing essentially all energy options simultaneously (1973–1980) to embracing "minimum interference" (1981–1988). Neither approach commends itself to the next president.

But how to change course effectively? How do we choose between an army of potential claimants on federal resources, all desperate for attention after nearly a decade's neglect? How can we separate likely winners from losers without wasting large sums that we don't have?

In a federal system, promising answers to questions like that can sometimes be found far afield from Washington, D.C. In the course of a long-standing regulatory relationship, many states and electric utilities have jointly acquired extensive experience in minimizing the cost of meeting future electric power needs. The functional equivalence of conserved and produced power has not gone unnoticed, and techniques for evaluating ways to do both have evolved under the rubric of "least-cost energy planning."

What this term implies to its now numerous practitioners is that the public should not be called upon to invest in new energy production until its representatives have explored and exploited less costly energy savings. If a coal-fired power plant is in prospect, least-cost planners investigate whether efficiency improvements in residences, commercial buildings, and industries could meet the same needs at a lower cost per kilowatt-hour delivered. Costs are evaluated over the anticipated life cycles of the competing conservation and power plant options, from design through construction and operation to retirement.

It is not enough simply to identify hypothetical savings, of course; means must be found to convert opportunities into tangible efficiency improvements. The last decade has produced a wealth of concrete experience with the use of efficiency standards and incentive payments to acquire cost-effective conservation. Rather than paying one group of people to produce power, utilities increasingly are paying a much larger group of people to save it.

Initial results from these initiatives have been dramatic; the nation's first officially adopted least-cost plan indefinitely deferred all new large-scale generators in the Pacific Northwest, which earlier had been home to one of the world's most ambitious nuclear power plant construction programs.[41] Utilities in the Northwest and California were spending almost half a billion dollars annually on conservation "resources" in the mid-1980s, and by 1986 their regulators had established savings targets exceeding the peak power production of 36 giant coal-fired power plants.[42] A national survey in 1987 determined that "between one-third and one-half [of] the utilities in the country are now offering energy efficiency rebate programs," which provided ratepayer-funded cash rewards for customers' decisions to save electricity.[43] That same year, least-cost planning initiatives were identified in at least 37 states.[44]

For purposes of such initiatives, minimizing society's costs means more than just comparing the projected dollar outlays for various resources, including conservation. Also important are considerations of flexibility: how long does a resource take to build? How susceptible is it to mid-course corrections if conditions change? Can it be built up gradually in small amounts, or does it create an enormous "all or nothing" proposition? In an era of rapidly changing economic conditions, these considerations carry growing weight. And at least some least-cost planners have recognized that environmental costs and benefits also belong in the balance.

At the national level, these analytical techniques provide the most promising ways to develop energy policies capable of winning broad support. What we need now is not prefabricated energy policy tailored to someone's predispositions, but a fair process that can build a consensus around priorities for meeting national energy and environmental goals at minimum cost. A least-cost planning inquiry would allow winners and losers to emerge on the merits. Conservation and supply options would compete on equal terms for access to public resources.

This requires a prompt and thorough inventory of our conservation and supply options, coupled with a comparison of their expected costs. For example, as the federal government weighs oil and gas drilling in frontier locations, it would seem only logical to investigate whether the nation could expect equivalent or larger returns by improving the fuel efficiencies of its vehicles, equipment, and housing. If so, which option delivers a reliable unit of energy at the lowest cost, taking risk-related and environmental issues into account, and how much can it deliver? The resulting rankings should guide allocation of federal agencies' regulatory and budgetary resources.

It is not difficult to demonstrate that conservation options *could* fare well in such an evaluation. We have demonstrated in detail elsewhere that improvements in the design of automobiles, houses, appliances, and mass transit systems could plausibly deliver at least twice as much oil and oil substitutes as the Arctic National Wildlife Refuge plus all undeveloped undersea lands under federal control.[45] These lands are a major source of controversy between environmental and energy interests. Before determining their fate, the federal government owes the nation a careful assessment of the magnitude and cost of conservation alternatives, including but not limited to the restricted group that we have been able to canvas.

These same "least-cost" techniques would allow the administration and Congress to set priorities among proposed solutions for the nation's most pressing environmental and security problems. The best way to reduce the coal combustion that promotes acid rain and global warming is not necessarily to build power plants that use other fuels; we might save more coal at less expense and risk by investing in more efficient buildings and industries. Must we revive nuclear power in order to battle the "greenhouse effect," or can our energy-service dollars go further by pursuing other ways to avoid burning fossil fuels? Are drilling rigs in the Arctic National Wildlife Refuge really among the most promising antidotes for dependence on Persian Gulf oil? Again, the federal government will not know, and should not act, until it looks.

In that process, the unfocused experimentation of the 1970s should pay some unexpected dividends. Thanks to many an uncoordinated demonstration project, the national energy archives hold extensive data on many of the technologies that will figure prominently in the proposed least-cost assessment. Other invaluable information will be available from state regulators and utilities.

Experience elsewhere suggests that a national least-cost energy plan could readily emerge within a year of the 1989 Inaugural. While many advocates of such a plan believe that conservation would fare well in the final version, others have very different expectations. Some still insist that nuclear power is poised for recovery, and we do not lack for enthusiasts about sun, wind, and tides. The point is to stage a fair competition in a public forum. We may not wind up cavorting all over the universe, but we will greatly improve the odds against destroying ourselves.

In Brief: Whatever Happened to Alternative Energy Resources?

Solar Heating and Cooling Systems: Installed in less than 1 million of the nation's 80 million households, and rapidly losing momentum. From a peak of 120 domestic manufacturers, 100 firms recently went out of business, taking with them 30,000 jobs.[46]

Ocean Thermal Energy Conversion: No systems in commercial operation.

Tidal Power: No systems in commercial operation.

Electricity from Industrial Processes' Waste Heat: From 1978–1988, some 2800 of these "cogeneration" facilities enlarged U.S. generating capacity by about 4%.[47]

Wind Energy: Accounts for less than one-third of 1% of installed U.S. generating capacity.[48]

Electricity and Steam Heat from Geothermal Resources: Installed generating capacity is comparable to the totals for wind, with most of it originating in a single California field.[49] Steam heat from geothermal resources is not a significant factor outside the scenic but constrained environs of Boise, Idaho.

Small-Scale Hydropower: We have about 1,500 of these units, which enlarge national generating capacity by about 1%.[50]

Wood-Fired Power Plants: U.S. utilities have built four of them. Total capacity is less than one-thirtieth that of the small-scale hydropower inventory.[51]

Solar Thermal Generators: We have eight systems, all of them in California, and their combined capacity does not even match that of the wood-fired units.[52]

Photovoltaics: The international industry's worldwide sales in 1986 were 23 megawatts, about half the capacity of the smallest U.S. wood-fired power plant.[53]

Alcohol-Based Fuels: Alcohol additives for gasoline exceeded 1.7 billion gallons in 1987, but still represented only about 1% of the energy content of the gasoline consumed during that year by U.S. vehicles.[54] In January 1988, the U.S. Department of Energy counted about 1,000 alcohol-fueled vehicles in the United States (out of at least 130 million cars and light trucks).[55] An abandoned Louisiana refinery for a new generation of alcohol fuels had cost the United States Treasury $70 million as of June 1988; the facility never opened following its completion in 1986.[56]

Synthetic Fuels: Energy supply from synfuels plants is less than one-sixth the output of small hydroelectric units, notwithstanding continued operation of the mammoth Great Plains Coal Gasification Unit (which defaulted on more than $1.5 billion in federally guaranteed loans).[57]

Fusion Power: Technology still under development; no systems in or near commercial operation.

Notes

1. The authors gratefully acknowledge the encouragement and financial support of The Florence and John Schumann Foundation.
2. National Appliance Energy Conservation Acts of 1987 and 1988. (Savings reflect only the impact of the initial standards, which are subject to substantial improvements over time.)
3. *Final Appropriations for Selected Energy, Defense Programs in FY-88,* Inside Energy/with Federal Lands, 5–6 (January 4, 1988).
4. Congressional Research Service, Renewable Energy: Federal Program and Congressional Interest, Tables 2 and 3 (March 2, 1988) (percentage declines reflect inflation-adjusted spending).
5. See R. Watson, *Oil and Conservation Resources Fact Sheet: A Least-Cost Planning Perspective* (Natural Resources Defense Council: July 1988).
6. The most popular 1940s model was a 7 cubic foot, manual defrost unit that consumed about 350 kWh/year, compared with a 17–18 cubic foot, automatic defrost unit for the mid-1970s, which averaged about 1750 kWh/year.
7. U.S. Energy Information Administration, 1986 Annual Energy Review, 59.
8. The actual percentages were 144% (energy) and 42% (population).
9. Scenic Hudson Preservation Conference v. FPC, 354 F.2d 608 (2d Cir. 1965), cert. denied, 384 U.S. 941 (1966).
10. 42 U.S.C.A. § 5901(a).
11. 42 U.S.C.A. § 5901(c).
12. Federal Nonnuclear Energy Research and Development Act of 1974, 42 U.S.C.A. § 5905(a).
13. See Natural Resources Defense Council, *Choosing an Electric Energy Future for the Pacific Northwest: An Alternative Scenario,* 123 (1980).
14. The remark appears in an April 16, 1974, energy policy speech entitled "The Year 2000 Revisited."
15. U.S. Department of Energy, *Domestic Policy Review of Solar Energy,* iii (February 1979).
16. C. Shea, *Renewable Energy: Today's Contribution, Tomorrow's Promise,* 37 (Worldwatch Paper 81, January 1988).
17. S. Diamond, *"Synthetic Fuel Plant Scuttled,"* New York Times, August 2, 1985; U.S. General Accounting Office, *Synthetic Fuels: Status of the Great Plains Coal Gasification Project,* 5 (November 1987).
18. Congressional Research Service, note 4 above.
19. U.S. Department of Energy, *The National Energy Policy Plan,* 1 (October 1983).
20. Idaho and California figures were reported by utilities in response to a Senate inquiry; see generally Implementation of the Public Utility Regulatory Poli-

cies Act, Hearings Before the Senate Committee on Energy and Natural Resources, 99th Cong. 2d Sess. 58, 114 (1986). Figures for Maine are from Central Maine Power Co., Cogeneration/Small Power Production: Pre-Decrement Through Subsequent Decrement (1986).

21. U.S. Energy Information Administration, *Monthly Energy Review,* 40 (January 1988).

22. These losses are documented in Cavanagh, *Least-Cost Planning Imperatives for Electric Utilities and Their Regulators,* Ten Harvard Environmental Law Review 299, 302 & n. 11 (1986).

23. L. Brown, *State of the World 1986,* 100 (Worldwatch Institute, 1986).

24. Testimony of Donna R. Fitzpatrick, Assistant Secretary for Conservation and Renewable Energy, before the Committee on Energy and Commerce, Subcommittee on Energy and Power of the U.S. House of Representatives, 1 (December 2, 1987).

25. The calculations assume that consumption would have reached 103 quadrillion BTUs ("quads"), exceeding domestic production actually recorded for 1987 by 38 quads. Actual imports peaked at 26% in 1977. See U.S. Energy Information Administration, note 20 above, 3.

26. U.S. Energy Information Administration, *Monthly Energy Review,* 3–4 (January 1988) (import shares for 1973 and 1987 were 19% and 18%, respectively).

27. *Id.*

28. Rosenfeld and Hafemeister, "Energy-Efficient Buildings," 258 *Scientific American* 78 (April 1988).

29. Data reported in this paragraph were supplied by Dr. David Goldstein, Senior Staff Scientist, Natural Resources Defense Council.

30. The totals assume 500 megawatts per power plant.

31. Rand Corporation, *California's Electricity Quandary,* Vol. III, vi (September, 1972), prepared for the California State Assembly (R-116-NSF/CSA).

32. Rosenfeld and Hafemeister, note 28 above, 84.

33. J. Goldemberg et al., *Energy for a Sustainable World,* 63–67 (World Resources Institute: September 1987).

34. *Id.,* 65.

35. See *id.,* 64 (list of prototypes seating up to 6 passengers with fuel economies up to 98 miles per gallon, built by Volkswagen, Volvo, Renault, Toyota, and Cummins/NASA); M. Renner, *Rethinking the Role of the Automobile* (Worldwatch Institute: June 1988). (Renault has reached 124 MPG in testing the "Vesta" prototype; Peugeot's "ECO 2000" is another 70–100 MPG model.)

36. A. Lovins and H. Lovins, *Drill Rigs and Battleships Are the Answer (But What Was the Question?),* 7 n. 30 (April 1988) ("during 1978–86, Japan's GNP grew 36% while primary energy consumption grew only 4% (compared with 19% and −5% for the U.S.); Japan, even though more efficient to start with,

outpaced the U.S. by 4:3 in efficiency gains"); Rosenfeld and Hafemeister, note 28 above, (if U.S. were as efficient as Japan, "this country would consume half as much energy as it does today and save $220 billion per year").

37. D. Tirpak, *Energy Efficiency and Structural Change in the Industrial Countries: Implications for the Greenhouse Problem,* 4 (Office of Policy Analysis, U.S. Environmental Protection Agency, May 1988).

38. National Research Council, *Responding to sea level Rise* (1987); United Nations Environment Program, *Report of the International Conference on the Assessment of the Role of Carbon Dioxide and Other Greenhouse Gases in Climate Variations and Associated Impacts* (Villach, Austria, Oct., 9–15, 1985).

39. World Resources Institute, *A Matter of Degrees: The Potential for Controlling the Greenhouse Effect,* 18, 21, 22 (April 1987) (calling for acceleration of current annual rate of overall efficiency increases from 0.2% per year to 1.5% per year).

40. Brockman, "Acid Rain: Corroding United States–Canadian Relations," 6 *Journal of Energy Law and Policy,* 357, 361 (1985).

41. The plan in question was developed by the Northwest Power Planning Council for Idaho, Montana, Oregon, and Washington.

42. California's eight investor-owned utilities spent almost $1 billion on conservation programs between 1984 and 1986, according to the Energy Conservation Program Summaries published for those years by the Evaluation and Compliance Division of the California Public Utilities Division. By 2005, savings from programs already adopted in California are expected to reach 12,858 megawatts. California Energy Commission, Conservation Report II–, 11. The Northwest Power Planning Council estimates that utilities in Idaho, Montana, Oregon, and Washington invested between $800 and $900 million on conservation resources between 1981 and mid-1987. Northwest Power Planning Council, *A Review of Conservation Costs and Benefits,* 2 (October 1, 1987). The council has identified ten programs that are expected to save 3900 average megawatts—the equivalent of 5,300 peak megawatts operating at 70% capacity—if load growth is at the high end of the range deemed plausible over the next two decades. Northwest Power Planning Council, *1986 Northwest Conservation and Electric Power Plan,* 8–3 (1986).

43. "Study Finds Increasing Use of Rebate Programs as Utilities Seek Alternatives to Generation," *Energy Conservation Digest* 1, (January 4, 1988) (summarizing survey published by Electric Power Research Institute).

44. P. Markowitz, L. Speer, and N. Hirsh, *A Brighter Future: State Actions in Least-Cost Electrical Planning,* 1 (Energy Conservation Coalition: 1987).

45. The details of the calculation appear in R. Watson, note 5 above.

46. Data on number of systems from U.S. Department of Energy, *Energy Security,*

205 (March 1987); data on firms from Testimony of Scott Sklar, Director of the Solar Energy Industries' Association, before the House Subcommittee on Energy Research and Development of the Committee on Science, Space and Technology, March 11, 1987.

47. L. Hobart, *Impressions Can Mislead,* Public Power (Nov./Dec. 1987), 5–6.

48. U.S. Department of Energy, *Energy Security: A Report to the President of the United States,* 205 (March 1987) (1400 MW had been installed by end of 1986); C. Shea, note 16 above, 38 (projecting 1460 MW for California, where virtually all wind capacity is located, by end of 1987).

49. *Energy Security,* note 48 above, 203 (2000 MW).

50. *Id.* 206 (1460 plants, 6800 MW).

51. C. Shea, note 16 above, 21 (214 MW).

52. *Id.,* 30 (178 MW).

53. *Energy Security,* note 48 above, 204.

54. Based on data compiled by Chris Calwell, Natural Resources Defense Council, from various U.S. Department of Energy sources.

55. U.S. Department of Energy, *Assessment of Costs and Benefits of Flexible and Alternative Fuel Use in the U.S. Transportation Sector,* Appendix D (January 1988).

56. K. Schneider, "An Ethanol Fuel Plant That Never Opened," *New York Times,* June 16, 1988, 8.

57. U.S. General Accounting Office, *Status of the Great Plains Coal Gasification Project,* (November 1987).

Chapter 20

AN ANALYSIS OF THE MONTREAL PROTOCOL ON SUBSTANCES THAT DEPLETE THE OZONE LAYER

Office of Technology Assessment

INTRODUCTION AND SUMMARY

A conference on the Protection of the Ozone Layer was convened by the United Nations Environment Program (UNEP) from March 18–22, 1985. The conference adopted the Vienna Convention for the Protection of the Ozone Layer and a "Resolution on a Protocol Concerning Chlorofluorocarbons." The resolution called for the development of a protocol "that addresses both short and long-term strategies to control equitably

From a staff paper prepared by the Oceans and Environment Program, Office of Technology Assessment, U.S. Congress. The views expressed in this staff paper do not necessarily represent the views of the Technology Assessment Board, the Technology Assessment Advisory Council, or individual members thereof.

global production, emissions, and use of chlorofluorocarbons (CFCs), taking into account the particular situation of developing countries as well as updated scientific and economic research." The United States ratified the convention in August 1986.

Pursuant to the convention and the resolution, the executive director of UNEP convened a diplomatic conference to adopt such a protocol.[1] Elements of many different proposals for limiting the emissions of ozone-depleting compounds were ultimately incorporated into the Montreal Protocol. The document, drafted and signed on September 16, 1987, seeks to inhibit production, consumption, and trade of some of these compounds,[2] while protecting the key interests of participants in the conference and encouraging all countries to become parties to it; the text reflects these differing and sometimes conflicting goals. In some instances, difficult issues were unresolved and key terms defined ambiguously, allowing for various interpretations. For example, developing countries are allowed to increase production of CFCs and halons to meet "basic domestic needs"; it is unclear whether these countries may export products made with CFCs and halons under this terminology or not.

It is not possible, therefore, to precisely describe the protocol and to exactly define its consequences on the emissions of ozone-depleting compounds. However, it is possible to summarize the language of the agreement, lay out a timetable for its provisions, identify principal uncertainties, and discuss in a general manner the impact of the protocol on the production of ozone-depleting compounds.

The protocol has been signed by 24 countries to date; almost all are developed countries, which also tend to be the major producers and consumers of the ozone-depleting compounds. Additional countries, especially the USSR and Australia, are very likely to sign; for most other countries, however, the likelihood of signing is unknown. If the protocol is signed and ratified quickly by the requisite share of producers (a minimum of 11 parties sign, representing at least two-thirds of global consumption), and if the Vienna Convention is ratified, it will have entered into force on January 1, 1989. Six months later, provisions designed to limit the production and consumption of specific compounds go into effect. The compounds are divided into two groups of "controlled substances," Group I (certain CFCs) and Group II compounds (specific halons), each subject to different limitations.

Central to the protocol is the distinction it makes between two groups of countries: (1) countries with relatively high levels of consumption of the controlled ozone-depleting substances, and (2) developing countries with relatively low levels of consumption. The requirements for inclusion in the latter group are provided in Article 5 of the protocol. (Those countries are henceforth referred to as "Article 5 countries"; the other countries will be termed the "developed countries.")

The principal difference between the developed and the Article 5 countries is in the timing of the production and consumption limitations. Beginning in mid-1989, the developed countries must freeze production and consumption at 1986 levels. Group I compounds must be cut to 50% of 1986 levels over the next 10 years; Group II substances may remain at 1986 levels. The Article 5 countries are given a 10-year delay (beginning in 1989) during which they are free to increase production and consumption within certain limits. Then, they too must cut production and consumption of Group I compounds over a 10-year period and freeze consumption and production of Group II compounds. But their restrictions will be pegged to the consumption levels during 1995–1997, rather than to the levels of 1986.

The Montreal Protocol can significantly inhibit the worldwide growth in the consumption of the compounds that deplete the layer of stratospheric ozone around the earth. OTA estimates that world use of the two major Group I chlorofluorocarbons, CFC-11 and CFC-12, could be more than twice current levels by the year 2009, if growth continued unabated (see Figure 20.1, Scenario D). With the protocol, production of ozone-threatening substances will be considerably less than it would be in the absence of any international restrictions.

However, the general perception[3] that the protocol will achieve a 50% reduction in the production of controlled compounds by the year 1999 appears incorrect. There are several key reasons why production and consumption probably will not be cut by 50% by 1999, even under very optimistic conditions. First, only developed parties must reduce consumption levels by 50% by 1999; growth in consumption of the controlled compounds is permitted among most of the world's developing countries until the end of the century. Second, production and consumption of Group II compounds are not cut back under the protocol. They are frozen at 1986 production and consumption levels for developed countries but allowed to grow until mid-1999 for Article 5 countries at which point they are frozen at the average of 1995–1997 production levels. Third, the Soviet Union is permitted to increase its production and consumption by two-thirds before cutting back.

By 2009, if the protocol is implemented, adhered to, and if developing countries do not increase exports of products made with controlled substances during the 10 years their consumption and production is allowed to grow, OTA estimates consumption of CFC-11 and CFC-12 could range from a 20% *increase* to a 45% *decrease* from 1986 levels. There is uncertainty, however, about the magnitude and direction of the actual change in CFCs and halons that will occur under the protocol. This uncertainty stems from several unknowns:

1. Incomplete information on production, consumption, and trade of (a) CFC and halon compounds and (b) products made with these substances. For example, baseline information, which is necessary to calculate expected reductions under the protocol, is not presently available for most countries.
2. The number of countries that ultimately ratify the protocol.
3. The extent to which the parties comply with the provisions of the protocol.
4. The rates of growth in consumption of controlled substances in Article 5 countries during the 10-year grace period. The average levels achieved by the 1995–1997 period will be the critical reference number for calculation of later reductions.
5. The degree to which additional control provisions will be added to the protocol. For example, what will be the extent of the lists of products *made with* or *containing* ozone-depleting substances drawn up in accordance with Article 4? Control of imports from nonparties will be based on these annexes. The extent to which parties agree to these lists is also important; objectors need not ban imports of the products.
6. The degree to which other ozone-depleting substances not covered by the protocol may grow in response to the protocol.
7. The degree to which production and consumption drop further than is required by the protocol. This could result from widespread shifts in consumer preferences and/or rapid development of chemical substitutes that replace substantial portions of the controlled substance market.

OTA ANALYSIS

Given the uncertainties associated with the protocol itself, today's data base, the number and behavior of parties, and future worldwide economic activity, estimates of production and consumption of CFCs and halons in future years are necessarily uncertain. It is difficult, therefore, to forecast reductions that might result from it by the year 2009—the year by which all mandated reductions will have taken place. Scenarios can be constructed to provide reasonable upper and lower bounds. A large consumption-cutback scenario, with a maximum reduction in use of ozone-depleting substances, is one where every nation in the world abides by the protocol. A low consumption-cutback scenario would include only current signatories as subject to the requirements of the protocol. While it is possible that the quantity of controlled substances could exceed these extremes, this range provides a plausible estimate of the bounds of the protocol.

OTA analyzed the potential effects of the treaty in two ways. First, as summarized in Figure 20.1, we examined the sensitivity of the ex-

FIGURE 20.1
Consumption Scenarios: CFC-11 and CFC-12

	Developed (thousand metric tons)	Developing (thousand metric tons)	Total (thousand metric tons)	Change from 1986 levels
1986	550–660	120–230	700–890	
SCENARIO A: *The Whole World Signs the Treaty*				
1999	280–330	190–380	490–710	−15% to −35%
2009	280–330	90–190	390–520	−40% to −45%
SCENARIO B: *The World Signs the Protocol, Minus Some Key Countries*[a]				
1999	290–350	190–380	510–720	−15% to −30%
2009	310–370	190–370	530–740	−15% to −30%
SCENARIO C: *Current Protocol Signatories*[b]				
1999	300–360	190–380	520–740	−15% to −30%
2009	330–400	300–600	670–1000	−10% to +20%
SCENARIO D: *The Treaty Never Goes into Effect*				
1999	770–920	190–380	1000–1300	+40% to +60%
2009	1090–1310	330–650	1500–1960	+110% to +140%

Note: Estimates represent upper and lower bounds across a range of eight simulations per scenario; thus, numbers may not add across rows.

[a]Much of the growth in CFC consumption could occur in a fairly small set of developing countries: China and India—poor countries with enormous populations and reasonably optimistic economic prospects; Indonesia, Brazil, and Mexico—with GNP now in the middle range among nations; and Saudi Arabia, Iran, and South Korea—which have high levels of GNP per capita. (Kohler, Haaga, and Camm, "Projections of Consumption of Products Using Chlorofluorocarbons in Developing Countries," January 1987. Because Mexico has already signed the protocol, we do not include it in this category).

[b]Signatories plus U.S.S.R. and Australia.

pected changes in consumption of controlled substances to changing numbers of parties to the treaty. These scenarios look only at changes in chlorofluorocarbons CFC-11 and CFC-12 because disaggregated data are available only for these compounds. However, CFC-11 and CFC-12 combined represent about 77% of world use of the substances controlled under the protocol. Where detailed country-by-country numbers for consumption do not exist, use[4] is estimated based on GNP. EPA notes that there is a consistent relationship between billions of dollars of Gross National Product ($ billions GNP) and metric tons of CFCs consumed (see Figure 20.2). The range is between 40 and 80 metric tons per billion dollars GNP, with the average about 60 metric tons per

FIGURE 20.2

Production/Use per Billion U.S. Dollar GNP for Various Countries for CFC-11 and CFC-12

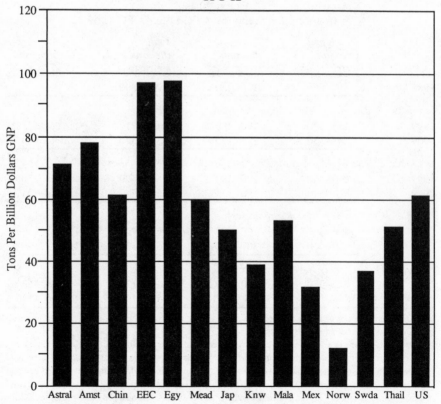

(*Source: "Chlorofluorocarbon Production and Use Data," assembled by ICF Inc. for the Environmental Protection Agency, February 1987.*)

billion dollars GNP. This relationship appears to hold for developing (Article 5) countries such as China as well as for developed countries like the United States. OTA used this relationship along with published statistics on current population, population growth rate per country, and rate of growth of per capita GNP to project CFC use in the future.[5] The ranges displayed in each scenario of Figure 20.1 bracket varying growth rates for the developed and developing countries.[6] The four scenarios shown here are:

- Scenario A: Large consumption cutback—all countries ratify the treaty. This results in worldwide reductions of 15–35% by 1999 and 40–45% by 2009 of CFC-11 and CFC-12 from an assumed 1986 baseline.
- Scenario B: All countries ratify the treaty except for current nonsignatories identified as pivotal to future CFC use (China, India, Indonesia, Brazil, Saudi Arabia, Iran, and South Korea).[7] This results in worldwide reductions of 15–30% by both 1999 and 2009 of CFC-11 and CFC-12 from an assumed 1986 baseline. By excluding only these eight countries from the protocol, much of the reduction in consumption of ozone-depleting substances possible under Scenario A is lost. This scenario demonstrates the importance of including key countries with large populations and/or favorable economic prospects in the treaty.
- Scenario C: Only the countries that have signed the protocol (plus the USSR and Australia) become parties. This results in worldwide reductions of 15–30% by 1999 and reductions of 10% to an increase of 20% by 2009 of CFC-11 and CFC-12 from an assumed 1986 baseline. This scenario demonstrates the importance of including developing countries in the treaty; most have not signed the protocol to date.
- Scenario D: The treaty never goes into effect. The world levels of CFC-11 and CFC-12 increase 40–60% over 1986 levels by 1999 and more than double by 2009. This scenario demonstrates the value of the treaty; even if only current signatories (plus the USSR and Australia) become parties as in Scenario C (above), consumption levels in the world will remain fairly close to 1986 levels.

In the second type of analysis, summarized in Figure 20.3, future consumption is calculated for *all* compounds covered in the treaty weighted by their ozone-depleting values.[8] Although we cannot examine the effect of adding or subtracting parties in this situation (because detailed data are not available for compounds other than CFC-11 and

FIGURE 20.3.

Projected Consumption of Controlled Substances: Large Consumption-Cutback Scenario (in thousands of metric tons)

	1986[1]		1995–97 Avg.[2] (Article 5 Parties Only)	1999		2009	
	Unweighted	Weighted[3]	Weighted[3]	Weighted[3]	% Change from '86	Weighted[3]	% Change from '86
DEVELOPED PARTIES[4]							
Group I	935	895	—	475	−47	475[5]	−47
CFC-11 and 12	740	740	—	370		370	
CFC-113, 114, 115	190	155	—	77		80	
Group II	15	94	—	94	0	94	0
Total Group I & II	950	990	—	570	−43	570[5]	−43
ARTICLE-5 PARTIES[6]							
Group I	180	180	265	265	+47	130	−26
FC-11 and 12	170	170	245	245		125	
CFC-113, 114, 115	14	12	21	21		10	
Group II	3	15	23	23	+53	23	+53
Total Group I & II	185	195	290	290	+49	155	−20

WORLD							
Group I	1,115	1,075	—	740	-31	605[5]	-44
CFC-11 and 12	910	910	—	615		495	
CFC-113, 114, 115	205	165	—	101		87	+7
Group II	20	110	—	118	+8	120	+7
Total Group I & II	1,130	1,185	—	858	-28	720[5]	-39

NOTE: All three-digit numbers are rounded to the nearest "5" (resulting in numbers which end with either "0" or "5").

KEY ASSUMPTIONS:
—The protocol is not altered.
—All countries become parties.
—Article-5 countries produce and consume controlled substances for products used only domestically.
—Production and consumption levels do not fall below maximum levels allowed by protocol.
—All of the Soviet capacity allowed under Article 2, paragraph 6, consists of Group I compounds; all of it is consumed domestically.

[1] The 1986 unweighted values are estimates provided in Exhibit 4–6 of United States, Environmental Protection Agency, Office of Air and Radiation, Office of Program Development, Stratospheric Protection Program, Draft Regulatory Impact Analysis: Protection of Stratospheric Ozone. Volume I: Regulatory Impact Analysis Document (Washington, D.C.: U.S. EPA, October 16, 1987).

[2] Based on the growth rates provided in Exhibit 4–5 of U.S. EPA.

[3] The weighted values are computed by multiplying the unweighted values by the "ozone depleting potentials" listed in Annex A of the Montreal Protocol.

[4] The category of "developed" countries includes the United States, the USSR and the Eastern Block Countries of Europe, and "other developed countries," as defined in U.S. EPA.

[5] This includes 50% of the allowance provided to the Soviet Union under Article 2, paragraph 6. Because the allowance is estimated to be 50,000 tons, 50% or 25,000 tons will be permitted by 2009.

[6] The category of "Article-5 countries" includes China and India, "Group I" Developing Countries and "Group II" Developing Countries as defined in U.S. EPA.

CFC-12) this situation is a useful check on Scenario A. We use aggregated consumption numbers by region of the world for Group I and Group II compounds, EPA growth rates to calculate future emissions, and apply the requirements of the treaty to the whole world. Figure 20.3 shows current and future production for developed countries, Article 5 (developing) countries, and the world. The results show that by the year 2009, if the whole world abides by the protocol, consumption of controlled substances could decrease by about 40% from 1986 levels in terms of the ozone-depletion potential. The range calculated in Scenario A for the same degree of world participation is 40–45%; in both analyses this is somewhat less than the 50% cut that the United Nations anticipates by the year 1999.[9]

Either type of analysis—using detailed country-by-country consumption numbers for chlorofluorocarbons CFC-11 and CFC-12 only Figure 20.1 or using large world regions but including all ozone-depleting substances covered by the protocol (Figure 20.3)—converges on the same conclusion. Even with world cooperation through the treaty, OTA's analyses suggest that total reduction of ozone-depleting compounds would be somewhat smaller and slower than previously believed by some observers.

Greater worldwide reduction in consumption of ozone-depleting substances than OTA calculated could occur if:

1. The provisions in the protocol are tightened.
2. Consumption drops more than is required by the protocol, which may occur if countries take unilateral actions directed towards that end or if widespread changes in consumer preferences occur. The latter would require rapid development and market infiltration of alternatives.
3. CFC and halon consumption in developing countries grows more slowly than the ranges assumed by EPA or OTA.
4. Nonparties reduce production and consumption in accordance with the protocol in order to export controlled substances and related products to parties.

Smaller worldwide reduction in consumption of ozone-depleting substances than OTA calculated could occur if:

1. Fewer countries become parties to the protocol than are expected under the different scenarios.

2. Developed countries who are parties to the protocol increase their imports of products made with controlled substances from nonparties or Article 5 countries.
3. Article 5 countries increase their consumption of CFCs and halons at a faster rate than EPA or OTA estimates.
4. There is a significant increase in the use of other ozone-depleting compounds (including methyl chloroform, carbon tetrachloride, and CFC-22) not covered by the protocol. While none of the major compounds omitted from the protocol are as damaging to the ozone layer as the controlled substances (see Figures 20.4 and 20.5), they do pose a small but growing ozone-depletion threat.

Many of the ozone-depleting substances contribute to the "greenhouse effect" or global warming as well (see Figure 20.6).[10] The substances covered by the Montreal Protocol are important contributors to both problems partly because of their long residence times in the atmosphere. CFC-22 is an example of a chlorofluorocarbon that is not currently covered under the protocol. Its ozone-depletion potential is approximately 5% of that associated with CFC-12; it would take 20 units of CFC-22 to have the same effect on ozone as 1 unit of CFC-12. Its contribution to global warming is approximately one-tenth that of

FIGURE 20.4
Ozone Depletion Potentials per Molecule

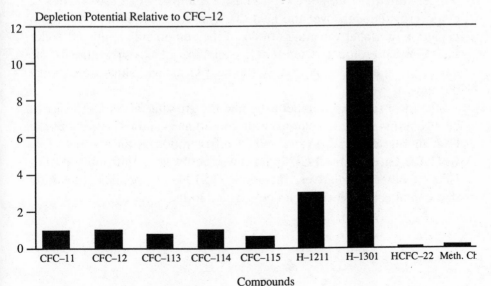

Depletion Potential Relative to CFC–12

Compounds

(*Source: Derived from EPA, 1987.*)

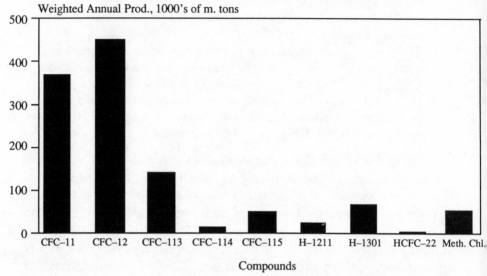

FIGURE 20.5.
Ozone Depletion Potentials per 1985 Worldwide Production

(*Sources: Derived from EPA, 1986 and 1987; and Hammitt et al., 1986.*)

CFC-12; it would take 10 units of CFC-22 to have the same impact on temperature that 1 unit of CFC-12 has. It is unlikely that use of CFC-22 could grow enough over the next 20 years to equal either the ozone-depletion or global warming effects of the compounds regulated under the Montreal Protocol. Therefore, use of CFC-22 as a substitute for the controlled substances, at least in the near term, will slow these global changes.

The most significant impetus behind the growing use of CFC-22 will be its continued use in stationary air-conditioners and its growing use in both mobile air-conditioners and in refrigeration as an alternative to CFC-12. Growth in CFC-22, however, could be significantly diminished if other alternatives, including FC-134a, a chemical substitute that contains no chlorine, can be adapted to these uses.

FIGURE 20.6
Ozone Depletion Potential and Greenhouse Effect for Selected Compounds

Compound	Ozone Depletion Potential	Greenhouse Potential
GROUP I		
CFC-11	1.0	0.3
CFC-12	1.0	1.0
CFC-113	0.8	0.3
CFC-114	1.0	**
CFC-115	0.6	**
GROUP II		
Halon 1211	3.0	**
Halon 1301	10.0	0.8
Halon 2402	6.0	**
COMPOUNDS NOT COVERED BY PROTOCOL		
Carbon Tetrachloride	1.06	**
HCFC-22	0.05	0.09
HCFC-123	**	0.01
HCFC-132b	**	0.01
HCFC-134a	0.0	0.04
Methyl Chloroform	0.1	**

General Note: For entries marked with **, there either is no known significant effect, or the data was not readily available at the time the analysis was conducted.
Notes for both Ozone Depletion and Greenhouse Potentials: Measured relative to CFC-12, which is set at 1.0.

(*Sources: Ozone depletion potential, U.S. Environmental Protection Agency,* Supplementary Information *provided with Proposed Rule on Protection of Stratospheric Ozone, December 1987; Greenhouse potential, E. I. duPont de Nemours & Company, written communication from Joseph M. Steed to OTA staff, dated January 1, 1988.*)

Notes

1. Participants and observers at the conference are listed in Appendix A of this analysis. [*Editor's Note:* Appendix A is not included here.]
2. The two types of ozone-depleting compounds limited under the protocol are chlorofluorocarbons (CFCs) as well as bromine-containing compounds (halons).
3. Telefax, 23 November 1987, from I. Rummel-Bulska, United Nations Environment Program, Nairobi, Kenya, "Response to inquiry from US Office of Technology Assessment regarding Montreal Protocol on Substances That

Deplete the Ozone Layer"; Environmental and Energy Study Conference Notice, "The Montreal Protocol on Substances That Deplete the Ozone Layer" 10/06/86; AP Newswire Story 11/20/87, "Stafford, Others, Urge Action on Ozone Layer"; AP Newswire Story 12/02/87, G. Darst, "EPA Plan Would Use Market Pressure to Control Ozone-Destroyers."

4. The protocol defines "consumption" strictly as *direct use* of the actual controlled CFC or halon, *not* in terms of the consumption or use of *products* made with controlled substances. Under the protocol's definition, a country manufacturing a refrigerator is the consumer of the controlled substance, even though the refrigerator ultimately may be used in another country. Because data based on the protocol's definition are not available, we base estimates of consumption on actual product use; only the ultimate user of the product is considered the consumer. Therefore, because developing countries are currently net importers of products made with or containing controlled substances, they appear in this analysis to have somewhat higher consumption levels than would be the case under the protocol's definition. Likewise, developed countries—net exporters of such products—appear to consume less.

5. World Population, rate of population increase, and GNP from *1987 World Population Data Sheet,* Population Reference Bureau, Inc.; average rate of growth of GNP/capita over the period 1970–1981, World Bank Statistics.

6. For developed countries we used 50–60 tons CFC per billion dollars GNP to project growth; for developing (Article 5) countries, this variable ranged from 40–80 tons CFC/$ billion GNP.

7. Kohler, D. F., Haaga, J., and F. Camm, *Projections of Consumption of Products Using Chlorofluorocarbons in Developing Countries* (Santa Monica: Rand Corp., January 1987), N-2458-EPA. This report included Mexico as an important country; since it has already signed the protocol, we did not include it as a nonsigner under this scenario.

8. The weightings are described in Annex A of the protocol:
Group I: CFC-11 (1), CFC-12 (1), CFC-113 (0.8), CFC-114 (1), CFC-115 (0.6).
Group II: Halon-1211 (3), Halon-1301 (10), Halon-2402 (to be determined).

9. Telefax, 23 November 1987, from I. Rummel-Bulska, United Nations Environment Program, Nairobi, Kenya, op. cit.

10. The greenhouse effect is caused by the buildup of carbon dioxide and other trace gases (such as chlorofluorocarbons, methane, and nitrous oxide) in the atmosphere. Currently, the amount of warming contributed by the trace gases is equal to that caused by carbon dioxide. (V. Ramanathan, Testimony of November 9, 1987, to the United States Senate, Committee on Energy and Natural Resources, Washington, D.C. 20510.)

Chapter 21

NEAR-TERM CONGRESSIONAL OPTIONS FOR RESPONDING TO GLOBAL CLIMATE CHANGE

William R. Moomaw

BACKGROUND

Two aspects of global change issues are striking to any observer. The first is the rapidity with which the problem is entering the public and political consciousness, and the second is the dramatic shortening of the projected time before the effects of climate change and sea-level rise begin to be felt. Public awareness is obviously a function of the closeness of the consequences, but what is remarkable is the desire to address what all of us recognize as the most challenging environmental, economic, and political issue we have ever faced. The complexity of

From a report published by the World Resources Institute, June 1988. An earlier version of this report was originally prepared for "The Congressional Staff Retreat on Climate Change" held at Airlie, Virginia, 10–11 March 1988.

human-caused global environmental change arises because of the intrinsic linkages that exist among the major issues considered here, the so-called greenhouse effect in which the emission of carbon dioxide and other gases released by energy production, agriculture, and industrial processes contributes to global warming. The warming and emissions in turn are linked to a host of related critical problems including climate modification, sea level rise, deforestation, acid rain, air pollution, and stratospheric ozone depletion. It would be more than a little presumptuous to assume that one could even identify all of the issues that Congress should address in a paper of this scope, much less describe all of the policy options that should be considered. Despite these reservations, I hope to provide a focus for the discussion of a congressional agenda that addresses near-term global change policy options.

Before examining specific proposals, it is important to state the context in which they are being offered. Despite the extensive work to date, many uncertainties remain about both the timing and the extent of global climate change. A pattern of global warming that has a distribution over the earth and atmosphere consistent with predictions of the greenhouse theory has been recently reported. Because the changes are slight, however, most scientists would like to observe a pattern of consistent effects before claiming that the human-induced greenhouse effect has been detailed. Some scientists predict that a definitive greenhouse fingerprint will be detected within a decade. The dilemma of waiting for conclusive proof in the face of such severe potential consequences is illustrated by the international Policy Issues Workshop held at Bellagio, Italy, which suggested that it might be thirty years before we would have definite proof that climate change was linked to our emission of greenhouse gases.[1] The report, released on June 6, 1988, also concluded, however, that a decade after confirming the greenhouse connection, we could expect a global temperature rise of 2.2° Fahrenheit, and an increase in sea level of three-quarters of a foot (middle range scenario). Because of numerous uncertainties, however, the Bellagio group estimated that by 2030 there was a 10% chance that warming could be less than 0.4°F with a slight decline in sea level or that temperature rise could exceed 6°F with a sea level rise of more than 3 feet. Even these dramatic figures, which imply global warming rates 10 to 100 times those that have ever occurred, and the warmest average global temperatures in the past 100,000 years, do not reveal the full extent of future warming commitment that will be realized a few decades later.[2]

To gain some insight into why the near future holds such dramatic changes in store when so little evidence of greenhouse gas induced global change is evident now, one needs to examine the history of the release of these gases. The greatest single contribution to global warming is carbon dioxide (CO_2) which is released in the combustion of fossil fuels. While CO_2 has been released to the atmosphere since the beginning of the industrial revolution nearly 200 years ago, two-thirds of all of the excess carbon dioxide has been added in the past 35 years.[3] If the projections are accurate, 35 years happens to be the approximate amount of time we have in the future before major consequences of greenhouse warming become substantial. Carbon dioxide and other greenhouse gases are certain to accumulate further during this period. Even were we to stop our additions today, we are committed to at least another 2.0°F of warming, and many decades must pass before carbon dioxide levels dropped sufficiently to reverse the temperature rise.[2]

We are therefore confronted by both the rate of change of global warming and its ultimate extent. Presumably, it will be easier to adapt to more gradual than to rapid change. Slow rates of change permit the introduction of new, climate-dependent infrastructure as regular replacement for old retiring facilities, and the development of new crops that are better suited to altered moisture and temperature levels. Indeed, the Bellagio group recommends that we set a goal of 0.1°C increase per decade as a management tool. This figure is based upon an estimate of the maximum rates of adaptation of natural ecosystems and of human societies to past responses to climate change. While slowing down emissions to reduce the rate of temperature increase makes great sense, this strategy appears to be risky and unworkable. There remain great uncertainties in our knowledge about the process of global warming, and we have an even poorer understanding of what a reasonable rate of increase should be. Finally, it seems an overwhelming task to create a fine-tuned global temperature control mechanism capable of responding to unforeseen natural events and the needs of the world's economy.

In addition to considering the rate of temperature increase, it is also necessary to determine whether there exists some upper level of change that is likely to be within the bounds of reasonable adaptability. Warming by 5.5°F, equivalent to that induced by a doubling of preindustrial atmospheric CO_2 to around 540 parts per million (ppm), is often used as a benchmark for comparison, but probably exceeds a warming level to which we might readily adapt. Were carbon dioxide the only greenhouse gas, such an increase would not be expected before the end of the next century. We now realize that other gases generated by human activity such as chlorofluorocarbons, methane, nitrous oxide, and tropospheric ozone currently contribute as much to global warming as carbon dioxide. With an observed rise of nearly 1.0°F during the past century,[4] and a commitment from all these gases to an even greater temperature increase from emissions already in the atmosphere, it is now possible that warming equivalent to a doubling of CO_2 could be realized during the decade between 2030 and 2040.[2] Without active intervention to reduce greenhouse gas emissions, temperature increases will, of course, continue their rise to catastrophic levels in the decades to follow.

The consequences of greenhouse warming are projected to be quite severe. The relatively modest rise in global average temperatures is magnified severalfold as one moves toward the poles. This may cause significant changes in weather and precipitation patterns with a major drying and warming of the present American grain belt during the growing season. The creation of desertlike precipitation patterns for the mid-continent would come just as we approach the depletion of the aquifer which underlies it. The implications for crops, forests, and other ecosystems are profound: the advantages of CO_2-stimulated plant growth offset by increased drought, forest fires, and species loss, a decline in forest and tundra ecosystems, and an expansion of grasslands and desert.[5] Few forests, for example, could move rapidly enough in response to current rates of temperature change with devastating consequences for many of our national forests, parks, and wildlife refuges. Existing irrigation and hydroelectric projects might no longer receive adequate rainfall, and many new projects may have to be built.

Finally, as indicated above, significant sea-level rise appears to be inevitable with serious implications for coastal populations, cities, water supplies, harbors, and estuaries. We can get some sense of the consequences of a sea-level rise of a few feet by observing the havoc created by similar rises in the levels of the Great Lakes and the Great Salt Lake. While sea-level rise would create strong incentives for urban renewal in some of our coastal cities, the cost of either moving them or building protective barriers would probably be in the range of hundreds of billions of dollars.[1] For example, a recent study estimated the cost for a modest-sized city like Charleston, South Carolina, to adapt to expected sea-level rise as $1.5 billion.[6] The Dutch have just completed the Deltaworks Project near the mouth of the Rhine which protects a few miles of coastline at a cost of over $5.0 billion. Globally, more than one-third of the population lives along the coasts and a sizable number would be adversely affected by rising sea level. EPA has apparently just begun a study of the economic costs of sea-level rise, but much more needs to be done to determine the cost of adaptive strategies and outright losses from global changes of this kind.

What factors might we keep in mind as we attempt to set priorities for congressional responses? The seriousness of the projected consequences and their time frames and degree of irreversibility provide important yardsticks against which we can measure various policy proposals. Despite the irreversible, adverse consequences of our release of greenhouse gases, the focus of concern appears to be more on the irreversible impact of policy changes that might be enacted to solve these problems. What if we are mistaken and there is no global warming? One way to approach this dilemma is to initially phase in those policies that are most easily reversed, that have the lowest negative impact on the current economy, are most effective in reducing greenhouse gas emissions, and which have additional environmental and economic benefits as well. In any case, there exist many problems which we should be addressing that happen to be linked to the greenhouse issue. By a proper choice of action, we can effectively address both a problem like stratospheric ozone depletion and global warming.

Considering these factors, we suggest that the following options be considered:

- First, build up support for research and long-term monitoring so as to have a sound basis on which to base medium and long-range policy.
- Second, ensure that the federal government is properly organized to respond to this complex problem that cuts across so many departments and agencies.
- Third, develop and implement international policies and strategies to deal with this global problem.
- Fourth, begin to implement policies that will slow the growth of greenhouse gases in the short term.
- Fifth, prepare strategies for adapting to unavoidable climate change.

A summary of proposed preventive and adaptive strategies organized by agency is presented later.

RESEARCH INITIATIVES

Our new awareness of the extent and timing of greenhouse effects has occurred as the result of new research tools and new strategies for carrying out "global science." Without the contribution from satellites, a worldwide network of monitoring stations, research laboratories, interdisciplinary institutes, and the array of regular and supercomputers to correlate the data and model the complexities of climate, we would remain blissfully ignorant of the threat we face. The obvious challenge to scientists is to develop more accurate predictive models and to devise a more sensitive test that can identify a greenhouse fingerprint or lack thereof as soon as possible.

To achieve those goals, there must be a sustained, decades-long scientific research effort which I suggest calling the "Global Change 2020 Program." Its goal would be to provide us with a clear vision of the greenhouse problem and its relationship to other serious issues before the year 2020 so that we can continue to employ cost-effective preventive policies and adaptive responses. The program would include our own domestic research, such as that outlined by the Earth Systems Sciences Committee of the NASA Advisory Council, and similar pro-grams at NSF and NOAA and the U.S. contribution to international projects through the International Geosphere Biosphere Program (IGBP), research coordinated by the United Nations Environment Pro-gram (UNEP), and various bilateral efforts such as those recently agreed to by the United States and the Soviet Union at the recent summit.

Congress will need to provide significant new funds to ensure that satellites for remote monitoring and experiments are available as are the computers needed to handle the monitoring data and to model the results. In addition to the need for new satellites like the polar platform, Congress should determine whether funding should be provided for equipment like Landsat that is slated for abandonment. Considerable research needs to be supported to understand the continuing increases in methane and nitrous oxide, two important greenhouse gases. The development of strategies to reduce the important air pollutant and greenhouse gas tropospheric (lower atmosphere) ozone should also be a high priority. Much more work needs to be done on climate dynamics, especially nonlinear effects. We also need to increase our understanding of the role of the biosphere in climate change, and the implications for a range of representative ecosystems utilizing both space and ground level observation. All aspects of global change research and monitoring are currently underfunded, even in the short term. Yet they could be expected to provide major insights into the greenhouse and other linked problems. Responsibility for particular programs must continue to evolve among NASA, NOAA, NSF, EPA, and DOE. Enhanced authority to coordinate and shape priorities and ensure cooperation among agencies, universities, the private sector, and international research efforts also needs to be provided to some office within the government.

A second area of needed research is the exploration of policy options. It may seem peculiar in a presentation of policy options to advocate research on the very proposals being advocated, but we need to learn which mechanisms are likely to be the most effective in reducing greenhouse gas emissions most rapidly, most completely, most economically, and with least disruption to society. Because the necessary actions will be so pervasive, it is essential that they have a high degree of support by the public in order to be both politically acceptable and capable of being implemented. The U.S. and other nations now have some experience with several strategies designed to reduce pollution and promote energy efficiency. There also exists a significant body of general research on the relative merits of tax breaks, gasoline or oil import consumption taxes, pollution (in this case carbon) taxes, incentives, trade-offs, penalties, regulations, and negotiations. What is needed is to apply each of these ideas—and any others one can think of such as the imaginative debt for tropical forest swap—to specific greenhouse problems in order to determine which approach is most effective in which circumstances. Carrying out policy research is perhaps the only possible way to resolve the ideological biases that engulf most policy debate.

ORGANIZING GOVERNMENTAL RESPONSE

The difficult challenge to policymakers is how to respond to such a supreme threat when the only evidence is a theory, the prediction of computer models, data revealing a slight warming over the past century, some tantalizing paleoclimate evidence, and a lot of uncertainty. If one is to give any credence to the predictions, then it is clear that some action is essential long before we can be certain of the extent of future damage and its timing. Waiting until we are certain that our own emissions are increasing global temperatures could well commit us to ruinous levels of climate change. Let me suggest that even though the greenhouse problem is more complex, it bears a superficial resemblance to national defense issues. In both cases we are confronted with possible, massive destructive consequences of uncertain extent at some unknown time in the future. Not wishing to risk the consequences of being inadequately prepared, our response to a possible military threat has been to develop defenses, never knowing for sure whether the weapons we build will stave off some future war. By contrast, we know for certain that we can prevent or postpone major induced climate changes by decreasing the release of greenhouse gases into the atmosphere. While I would hope we could develop less lethal defenses against climate change than we have against perceived military threats, I must agree with Stephen Schneider of the National Center for Atmospheric Research who has argued very persuasively that it would be a serious mistake not to treat the greenhouse threat as an equally serious strategic defense problem. To respond effectively will require mobilizing government authority and resources in new arrangements. Rather than create a new "Pentagon for global climate change," it is far more realistic to divide the many aspects of this problem among the appropriate departments and agencies and then to develop an effective means for coordinating them. In some ways we have already begun since a number of the scientific aspects are already coordinated by the National Climate Program.

For Congress there is the additional challenge of organizing the task of gathering the necessary information, developing and passing authorizing legislation, and then appropriating the necessary funding to implement the many facets of a complex response to a complex problem. A look at the various dimensions of the problem suggests that at least a dozen different subcommittee jurisdictions and as many federal agencies (not to mention state and local governments and the private sector) might be involved. Simply allocating tasks among the agencies and figuring out coordination mechanisms will be a major task. High-level positions will need to be created at the EPA and the Department of Energy, and coordinating offices will be necessary in the Departments of State, Interior, Agriculture, Housing and Urban Development, and Defense. To be certain that responsibilities are met in a coherent way requires that ultimate responsibility be vested in a single cabinet-level official with sufficient staff to coordinate the research and policy implementation.

There is one action that could be implemented quickly, would cost very little, requires no irreversible actions, and yet is effective in increasing awareness of the greenhouse problem. Congress should require that federal agencies consider the implication of their actions for global change issues such as the greenhouse effect, sea-level rise, stratospheric ozone depletion, and tropical deforestation as part of environmental impact statements (EIS) filed under NEPA. Congress may also want to require comments on global change when issuing new regulations or when reviewing certain continuing agency activities that might not normally be considered eligible for the EIS process.

INTERNATIONAL INITIATIVES

The recent achievement of the Montreal Protocol on Substances That Deplete the Ozone Layer has been hailed as evidence that a similar agreement might be reached with regard to limiting greenhouse gases. The two issues share in common a global dimension and the absolute need for global cooperation. It should be a high priority of the Congress to press for an international meeting within the next two years to develop an agreement for controlling greenhouse gases.

Secondly, we must increase our financial commitment and involvement in international research activities such as those sponsored by UNEP and the International Geosphere-Biosphere Program.[11] We also need to actively pursue bilateral research and policy options with the Soviet, Chinese, Japanese, and European governments. The recent summit agreement with the Soviets to cooperate on global climate change issues represents an important step forward.

Since much of the future increase of greenhouse gases will come from increased industrialization and agricultural production in developing countries, it is essential that development assistance projects sponsored by U.S. AID, the World Bank, and other development agencies promote technologies that protect global climate. This is particularly true for the energy field, where a recent study has shown the tremendous potential for the cost-effective introduction into developing countries of energy-efficient technology[9] that also reduces both capital expenditures and the debilitating cost of imported fuels that contribute so much to Third World debt.

SLOWING GROWTH OF GREENHOUSE GASES

ENERGY EFFICIENCY

Using less energy to provide the same end-use services not only reduces global warming through a reduction in carbon dioxide emissions, it also reduces the production of air pollutants that cause acid rain, decreases our dependence on foreign oil, improves our trade balance, and provides an opportunity for entrepreneurs to create new technologies. Improved efficiency also lowers the levels of another important greenhouse gas and smog component, tropospheric ozone. Congress went a long way in this direction by recently passing appliance efficiency standards (twice!), which according to one estimate will eliminate the need for 40 large coal-fired plants.[7] On the other hand, weakening the gas mileage standards for automobiles (CAFE standards) and allowing energy efficiency tax credits to expire has discouraged manufacturers and the public from choosing more efficient cars and homes. More efficient heating, lighting, and air conditioning of new buildings especially in the residential and commercial sector, as well as retrofitting existing structures, can have a dramatic effect on CO_2 emissions. Removing barriers to these innovations, introducing incentives, and considering compulsory national building standards like those in California should be high on the congressional agenda. Improving efficiency is the fastest and least-cost strategy for reducing CO_2 emissions for an economy like ours,[2, 8] and is also the most effective option for new energy projects in developing countries which will provide a large share of new CO_2 emissions in the coming decades.[9]

FUEL SWITCHING

Among fossil fuels themselves there are enormous differences in the amount of CO_2 emitted for each unit of energy produced. Coal releases about twice as much carbon dioxide as does natural gas, with petroleum approximately halfway in between. Refining and burning shale oil and other synfuels yield from 2.5 to 3.5 times as much CO_2 as does burning natural gas.[2] New gas turbine technology is so efficient that the effective emissions advantage over coal may climb from a factor of 2 to a factor of 3.[10] As the debate over the Clean Air Act and the support for "clean coal" technology indicates, replacing abundant U.S. coal with scarcer natural gas will not be politically easy. Furthermore, we and the Soviet Union have comparable coal reserves together accounting for half of the global total, while China and Europe each account for 10% more. Hence, we face the possibility of intense international disagreement as well as interstate rivalry over the use of coal.

Despite these obstacles, Congress did repeal the provisions of the Fuel Use Act which had prohibited the use of natural gas for electrical utilities and large industrial boilers. The provisions of PURPA also encourage smaller-scale production of electricity and cogeneration which is both more efficient and more readily accomplished using oil and gas rather than coal. Unfortunately, there still exist regulatory barriers to fuel switching including the possibility of co-burning coal with natural gas to achieve significant reductions in CO_2 and sulfur and nitrogen oxides at very low costs. Congress should amend the Clear Air Act to remove barriers to innovation, and ensure that efforts to meet local and national air quality standards do not exacerbate the greenhouse problem. For example, the use of methanol synthesized from coal or natural gas as a transportation fuel will increase CO_2 emissions. Similarly, the regulatory framework for utilities needs to be restructured to ensure that fuel-switching and energy efficiency initiatives that reduce greenhouse gas emissions of CO_2 and nitrous oxide as well as other pollutants are encouraged.

While it is neither politically possible nor practically desirable to close down existing coal-powered electric generators, it is not too soon to plan for their orderly replacement. As was pointed out by a congressional staff member at a strategic planning session on global climate change in January 1988, many coal plants will be retired within the next fifteen years, and most will be replaced in the next forty. A program of incentives to phase out the oldest, least efficient, and most polluting coal plants on an accelerated schedule would help with both the greenhouse problem and acid rain. A strategy similar to this was successfully pursued by the Tennessee Valley Authority during the 1970s. Natural gas can help bridge the energy supply route to the future along with reduced demand through efficiency improvements until renewable technologies such as solar and wind or an acceptable and economical nuclear option can be developed.

DEVELOP RENEWABLE ENERGY TECHNOLOGIES

Renewable resources are abundant, but often diffuse. In appropriate circumstances, however, they can make a significant contribution to energy supply. Electricity generation by wind power in the U.S. now exceeds 1,500 megawatts and is expanding. The U.S. geothermal capacity of 2,000 megawatts currently exceeds 40% of the world's total, and the immense Geysers geothermal plant in California may soon be joined by an additional 500-megawatt plant in Hawaii. Photovoltaic production of electricity continues to expand for remote sites as the cost of production drops with each new technological development; two companies are constructing photovoltaic power stations in the 10-megawatt range. Steam and electricity production from biomass and combustible solid waste is a growing industry. Communities are attempting to meet both their trash and energy needs (New York has declared this approach along with recycling to be a statewide policy) using new, efficient, less polluting technologies. Solar and wind technologies have the advantage of producing no greenhouse gases, while biomass and trash (principally paper and garbage) produce no more CO_2 than they consumed during growth of the raw materials that produced them.

It is clear that tax incentives offered by some states in combination with progressive utilities and utility commissions have strongly encouraged investment in renewables, whereas the termination of federal tax credits has virtually halted the boom in solar domestic hot water heating. A reconsideration of federal tax policies as they affect renewables is certainly in order at this time as is an examination of those utility policies that have been effective. A redistribution of federal research and development funds into several promising areas such as amorphous and crystalline silicon photovoltaics could yield large dividends in the near term.

REASSESS THE NUCLEAR OPTION

Despite its perceived and real problems, nuclear power does have the advantage of not producing any greenhouse gases in the production of electricity. The current leveling off of nuclear plant construction at less than 20% of U.S. electricity generating capacity is a consequence of excessively high capital costs, quality control problems, an inability to resolve the waste problem in a timely fashion, and a loss of public confidence in nuclear power following the accidents at Three Mile Island and Chernobyl. The rate of introduction of nuclear electric generating capacity is slowing in most parts of the world, and this appears to be a propitious time to assess this technology and its future. Congress may wish to address such questions as how much nuclear capacity do we need and how much are we capable of managing and operating safely? Should we begin research on a new generation of inherently safer and more economical designs? Given the significant cost advantage and ease of incremental introduction of energy-efficient technologies and the promise of renewables, how should we allocate our limited R&D funds between these options and nuclear power?

REFORESTATION

Not all of the carbon dioxide increase in the atmosphere arises from the burning of fossil fuels. The massive deforestation under way in the tropics and elsewhere is believed to contribute about one-fifth of the observed excess in the annual carbon budget. Halting the destruction of forests combined with reforestation will provide a significant sink for CO_2 and create additional benefits such as restoration of local ecosystems, reduce soil erosion, and establish sustainable wood fuel and timber resources especially in developing countries. In particular, we should reevaluate our foreign aid programs to minimize unnecessary forest destruction and to encourage reforestation. Perhaps development projects might carry compensating reforestation provisions to offset forest losses or to absorb the carbon dioxide released from industrialization projects. We also need to examine the international debt situation and initiate debt-for-forest swaps as well as find other strategies for reducing pressures on developing countries to further cut their forests. A Tropical Forest Action Plan has been prepared jointly by the World Resources Institute, the World Bank, the United Nations Development Program, and the Food and Agriculture Organization of the United Nations which describes various policy mechanisms that may be used to protect this resource and the economies of developing countries.[12]

REDUCE CHLOROFLUOROCARBON EMISSIONS

Chlorofluorocarbons (CFCs) now contribute 30 to 40% as much as CO_2 to global warming, and their release is increasing at nearly ten times the CO_2 rate. Their lifetime in the atmosphere is of the order of 100 years, and each molecule contributes a thousand times as much to global warming as does each molecule of CO_2. As is now well known, these industrial substances are strongly implicated in the destruction of the stratospheric ozone layer and the dramatic appearance of the Antarctic ozone hole. In 1976, the Congress reduced U.S. use of CFCs nearly in half by banning "nonessential" uses in most aerosol sprays while permitting their continued production for "essential" purposes such as refrigeration, air conditioning, and foam blowing agents. During the past decade, however, other uses have expanded, and companies have been unwilling to reduce production voluntarily. The recently signed Montreal Protocol commits the U.S. and other countries to additional significant reductions in CFC production. Encouraging the development of substitutes through tax incentives and negotiation with producers and by regulation that will hasten the replacement of these substances with environmentally less harmful chemicals would be one of the fastest and most effective ways to reduce future commitment to global warming while simultaneously slowing the rate of destruction of stratospheric ozone.

PRICING OPTIONS

Economists have long recognized the importance of prices in determining demand in market economies. More recently, their attention has turned to the problem of distorted prices that do not reflect the total costs of particular goods and services, and hence lead to a level of consumption that misallocates resources.

The most obvious of these false prices occurs whenever one particular commodity is subsidized with respect to another. While the United States has moved away from subsidies such as oil depletion allowances, which encouraged the inefficient use of petroleum, there still remain areas such as natural gas and electricity pricing and R&D funding that favor some energy options over others. It would be useful to examine all such subsidies to determine their effect on the economy and on environmental problems such as greenhouse gas emissions. Similarly, the cutting of tropical forests has been greatly accelerated by literally billions of dollars in subsidies, while our own U.S. Forest Service is losing at least $85 million annually on timber sales, according to a just-released study.[13]

A second pricing problem occurs when all of the real costs are not included in the price being paid by the user. The current low energy prices we are now enjoying do not include the costs of Persian Gulf security or environmental damage such as acid rain, smog, and the greenhouse effect. These artificially low prices encourage excessive consumption of nonrenewable energy resources along with their additional pollution burden. The simplest method for dealing with this problem is to place a fee on each form of energy that reflects its environmental or other social costs.

One option that would respond to the greenhouse issue is a carbon fee that would be highest for those fuels that emit large amounts of CO_2 for each unit of useful energy produced. Another often discussed proposal places a tax on imported oil to help dampen demand and raise revenues. Aside from the trade barrier questions this approach would create, this strategy would encourage us to "drain America first." A more effective approach would be to charge a fee on all petroleum products or perhaps only gasoline. As unpopular as gasoline taxes are, they have an advantage during this time of low fuel prices in more accurately reflecting future oil replacement costs. Not only would such a fee promote energy efficiency and reduce CO_2 and other emissions, it would also make our economy less vulnerable to the inevitable shocks we face in the near future as our domestic reserves continue to decline. Finally, as Federal Reserve Chairman Alan Greenspan has observed, a modest increase in the gasoline tax would make a significant contribution to reducing the federal deficit.

The imposition of fees also seems an attractive option for reducing emissions of chlorofluorocarbons by placing a sufficiently high fee on these industrial and commercial chemicals that both deplete the ozone layer and contribute significantly to the greenhouse effect. By raising the price sufficiently, we can encourage the rapid introduction of less harmful substitutes and the recycling and reuse of those CFCs that are difficult to replace. An alternative strategy being pursued by EPA is to ration the production of CFCs and let the price rise in response. Unfortunately, this approach allows large windfall profits to accrue to the producers and lowers incentives for the rapid replacement of harmful CFCs. Care must, of course, be taken to ensure that substitutes designed to protect the ozone layer also decrease greenhouse warming.

The basic focus of all of these suggestions is to use the market mechanisms in the economy to reduce emissions of greenhouse gases and other pollutants. An advantage of using pricing mechanisms is that the incentives can be adjusted to achieve any desired level of emissions reduction, and the emitter may determine the most cost-effective method. It may be necessary if energy prices are raised in this manner to respond to the needs of low-income citizens with some form of assistance.

PREPARING FOR ADAPTATION

Since we cannot respond rapidly enough to stop all future contributions to global warming, and because there is already a commitment to a significant global warming from the greenhouse gases that have already been released to the atmosphere, it is essential that we consider adaptation options now. The longer we wait to take action to slow the buildup of these gases, however, the more we will be placed in a reactive mode of coping with the consequences of our actions (or lack thereof). Comparing the cost of adaptive and preventive strategies is also useful in providing us with an informed basis for implementing the policy options that have been described in this paper. In the following paragraphs, several examples of adaptive strategies are described.

To protect future coastal development, *Housing and Urban Development* should revise the national flood insurance program to deny insurance to any dwelling constructed in an area that would be flooded by projected sea-level rise during its lifetime.

The *U.S. Department of Agriculture* should prepare a long-range plan for adapting U.S. agricultural and forestry policies to the potential consequences of climate change. This should include consideration of possible shifts in areas of agricultural and forestry production and major shifts in agricultural productivity and crop type. USDA and the Forest Service should also prepare for a large increase in the number and size of forest fires.

Water resource plans need to be prepared by the *U.S. Department of the Interior* and the *Army Corps of Engineers* to address the implications of potential changes in the timing, distribution, and variability of precipitation. This plan should include the need for water conservation policies, the capacity of urban reservoirs, revised estimates of dependable power from hydroelectric sites, predictability of river levels for barge traffic, and other implications of changes in precipitation for federal water projects and irrigation. DOI should also include climate change issues in its long-range management plans for national parks, wildlife reserves, and other public lands.

Strategies for managing public energy resources in an era constrained by water shortages, limits on coal development, and other potential results of global climate change should be developed jointly by the *Department of Energy* and the *Department of the Interior.* The implications of fuel switching, use of renewables, and improved efficiency on the demand for conventional fuels from public lands also need to be determined.

The *Federal Emergency Management Agency* should consider climate programs responsive to the potential for an increase in the frequency of severe storms, coastal flooding, and other weather extremes possible from global climate change. FEMA and *Health and Human Services* should study responses to increasing heat waves.

The *Department of State* needs to plan for increased aid requests from nations adversely affected by drought, desertification, and agricultural failures.

The *Department of Labor* should determine the implications for a significantly altered labor market should large-scale shifts in our pattern of energy use occur.

CONCLUSIONS

We clearly face a problem of global dimension that will affect all nations on a scale that has never been encountered in human history. The encouraging fact is that there is something that we can do about it. It is also clear, however, that we must begin acting now to slow the growth in the emission of greenhouse gases. Waiting until a clear greenhouse fingerprint is unequivocally proven will, if current models are correct, commit us to an additional warming.[2] These same studies also show, however, that by taking action of the kind described here, we can reduce future warming significantly.

The Congress is perhaps the most critical forum in which the course of the U.S. response to climate change will be decided. In times of tight budget constraints, priorities must be set for scientific and policy research as well as for programs within the Department of Energy and other agencies. Choices that will be made now will determine not only our own contribution to future warming, but will significantly influence the rapidly evolving international discussion of appropriate responses. A great deal can be accomplished in a cost-effective manner within the next few years by improving the efficiency of our transportation sector and both end-use and production efficiency of electric power. This strategy will have other, multiple benefits for clean air, balance of payments, and energy security as well. We can also move more rapidly to replace chlorofluorocarbons with chemicals that contribute less to both global warming and depletion of the ozone layer. The sooner we can begin reforestation programs both in the U.S. and in the tropics, the greater will be the CO_2 absorptive capacity during the critical first half of the next century. On the other hand, a commitment to a major synfuels program would accelerate the already rapid rise in global warming. By a careful choice of policies that simultaneously slow the release of greenhouse gases and help solve other problems, we can effectively and prudently buy ourselves some insurance against a rapid global warming and its destructive consequences.

PROPOSED AGENCY-BY-AGENCY ALLOCATION OF RESPONSIBILITIES

Environmental Protection Agency:

- Chair interagency committee and coordinate policy options studies
- Phase out emissions of CFCs
- Identify other opportunities to reduce trace gases
- Study impacts of climate change due to sea-level rise, and analyze the implications of warmer temperatures for human health, for air pollution, and for the natural environment
- Global tropospheric ozone policy

- Nitrous oxide (N_2O) strategy
- Global analysis of emissions
- Integration with existing air pollution programs
- Examine the consequences of sea-level rise for saltwater intrusion into coastal aquifers
- Study the consequences of sea-level rise on near-coast landfills, toxic waste sites, and sewage treatment plants

Department of Energy:

- Implement tighter efficiency standards on appliances and buildings to promote energy efficiency
- Phase out subsidies to publicly owned utilities and energy companies to reflect more accurately the cost of energy to consumers
- Revise natural gas and wholesale electricity pricing to reflect more accurately marginal costs and promote efficiency (FERC)
- Revise energy supply priorities in research programs to emphasize renewable energy technologies and new efficiency technologies
- Conduct policy analysis on global energy use and alternatives
- Analyze implications of climate change for the energy system (e.g., change in patterns of electricity consumption and effect of sea-level rise on energy facilities)
- Oversee energy efficiency goals for the federal government

State Department:

- Advocate climate initiatives at UN agencies
- Support bilateral and multilateral agreements on research and action to address the greenhouse problem
- Consider climate change issues in U.S. development assistance programs
- Prepare for impact of climate change on refugees and increased aid
- Explore debt swapping and other approaches as mechanisms to preserve tropical forests

Treasury Department:

- Support assessment of climate impacts in the lending process of the World Bank and other lending agencies supported by the U.S.
- Study carbon dioxide (CO_2) tax
- Study gasoline tax and other consumption-based taxes

Interior Department:

- Reform fossil fuel leasing policies
- Assess the implications of climate change for the management of national parks and other public lands
- Assess implications of rising sea level on coastlines (USGS) and consequences for water projects (Bureau of Reclamation)

Corps of Engineers:

- Determine consequence of sea level rise on harbors, coastal waterways, and beaches
- Evaluate implications of changed climate on water projects and navigable rivers

Agriculture Department:

- Study implications of climate change for agriculture and the priorities of agriculture research
- Study implications of climate change for management of the national forests including increased fires
- Study agricultural practices as sources of increased methane and nitrous oxide emissions
- Assist in development of a global reforestation strategy
- Analyze multiple stresses on crops and forests

Transportation Department:

- Improved standards on auto efficiency
- Assess impacts of climate change and sea level rise on transportation system

Defense Department:

- Assess implications of climate change for national security
- Reduce DOD energy use

Labor and Commerce Departments:

- Examine shift in labor market in energy, agriculture, and commerce as greenhouse strategies develop and implications for U.S. economy, employment, and trade

Council of Environmental Quality:

- Review greenhouse implications as part of the EIS process under NEPA

Coordinated science and policy research among U.S. Environmental Protection Agency, U.S. Department of Energy, U.S. Geological Survey, National Science Foundation, National Aeronautics and Space Administration, National Oceanic and Atmospheric Administration/ National Climate Program:

- Establish research goals for an improved understanding of global change and the earliest possible detection of a greenhouse fingerprint
- Develop long-range research programs to study and monitor global climate change over the next 30 years (Global Change 2020 Program)

References

1. Developing Policies for Responding to Climatic Changes Policy Issues Workshop. 1987. "Priorities for Future Management—a New Policy Issues Agenda." Bellagio, Italy.
2. I. Mintzer. 1987. *A Matter of Degrees: The Potential for Controlling the Greenhouse Effect.* World Resources Institute: Washington, D.C.
3. Department of Energy. 1988. "Draft Report on Carbon Dioxide and the Greenhouse Effect." Office of Basic Energy Sciences, Department of Carbon Dioxide Research Division, Department of Energy: Washington, D.C.
4. J. Hansen and S. Lebedeff. 1987. "Global Trends of Measured Surface Air Temperature." *Journal of Geophysical Research,* 92(D11): 13345–13372. American Geophysical Union: Washington, D.C.; J. Hansen and S. Lebedeff. 1988. "Global Surface Air Temperatures: Update Through 1987." Submitted to *Geophysical Research Letters* (1/12/88). American Geophysical Union: Washington, D.C.
5. United Nations Environment Program. 1987. *The Greenhouse Gases.* United Nations Environment Program: Nairobi, Kenya.
6. The following articles provide useful models for sea level rise impacts: J. Titus and M. Barth. 1984. "An Overview of the Causes and Effects of Sea Level Rise" in *Greenhouse Effect and Sea Level Rise,* p. 34 (Van Nostrand Reinhold Co., Inc.: New York, New York); T. Kana, J. Michel, M. Hayes, and J. Jensen. 1984. "The Physical Impact of Sea Level Rise in the Area of Charleston, South Carolina" in *Greenhouse Effect and Sea Level Rise,* p. 105 (Van Nostrand Reinhold Co., Inc.: New York, New York).
7. Natural Resources Defense Council. 1986. *Annual Report 1986–87.* Natural Resources Defense Council: Washington, D.C.
8. J. Goldemberg, T. Johansson, A. Reddy, and R. Williams. 1987. *Energy for a Sustainable World.* World Resources Institute: Washington, D.C.
9. J. Goldemberg, T. Johansson, A. Reddy, and R. Williams. 1987. *Energy for Development.* World Resources Institute: Washington, D.C.

10. E. Larson and R. Williams. 1987. "Steam Injected Gas Turbines." In *Journal of Engineering for Gas Turbines and Power, 109*:55. American Society of Mechanical Engineers: New York, New York

11. U.S. Committee for an International Geosphere-Biosphere Program, Commission on Physical Sciences, Mathematics, and Resources, National Research Council. 1986. "Global Change in the Geosphere-Biosphere Programme: Initial Priorities for an IGBP." National Academy Press: Washington, D.C.

12. Food and Agriculture Organization of the United Nations, United Nations Development Program, the World Bank, World Resources Institute. 1987. *The Tropical Forestry Action Plan.* FAO, UNDP, the World Bank, World Resources Institute: Rome, New York, and Washington.

13. R. Repetto. *The Forest for the Trees? Government Policies and the Misuse of Forest Resources* 1988. World Resources Institute, Washington, D.C.

FOR FURTHER READING

CARBON DIOXIDE AND THE CARBON CYCLE

Historic Records of CO_2 Increase in the Atmosphere

J. Barnola. "Vostok Ice Core Provides 160,000-Year Record of Atmospheric CO_2." *Nature 329* (1987): 408–414.

Historic Fossil Energy CO_2 Emissions and Available Fossil Energy Resources

R. Rotty and C. Masters. "Carbon Dioxide from Fossil Fuel Combustion: Trends, Resources, and Technological Implications." In *Atmospheric Carbon Dioxide and the Global Carbon Cycle*, J. Trabalka, ed. DOE.ER-0239. Washington, D.C.: U.S. Department of Energy, 1985.

Carbon Cycle Overviews

J. Olson et al. "The Natural Carbon Cycle." In *Atmospheric Carbon Dioxide and the Global Carbon Cycle*, J. Trabalka, ed. DOE.ER-0239. Washington, D.C.: U.S. Department of Energy, 1985.

G. M. Woodwell. "The Carbon Dioxide Question." *Scientific American 238* (1978): 35–43.

G. M. Woodwell. "Biotic Effects on the Concentrations of Atmospheric Carbon Dioxide: A Review and Projection." In *Changing Climate*, Carbon Dioxide Assessment Committee, National Research Council. Washington, D.C.: National Academy Press, 1983.

Non–Fossil Carbon Sources and the Response of the Biosphere

R. Detwiler and C. Hall. "Tropical Forests and the Global Carbon Cycle." *Science 239* (1988): 42–47.

Overview of Oceans as a Sink for CO$_2$

P. Brewer. "Carbon Dioxide and the Oceans." In *Changing Climate,* Carbon Dioxide Assessment Committee, National Research Council. Washington, D.C.: National Academy Press, 1983.

Oceanic Response to CO$_2$

T. Takahashi and A. Azevedo. "The Oceans as a CO$_2$ Reservoir." In *Interpretation of Climate and Photochemical Models, Ozone and Temperature Measurements,* R. Reck and J. Hummel, eds. New York: American Institute of Physics, 1982.

CO$_2$-Energy Models and Modeling

H. H. Rogner. "Long-Term Energy Projections and Novel Energy Systems." In *The Changing Carbon Cycle: A Global Analysis,* J. Trabalka and D. Reichle, eds. New York: Springer-Verlag, 1986.

R. Rotty. "Electrification: A Prescription for the Ills of Atmospheric CO$_2$." *Nuclear Science and Engineering 90* (1985): 467–474.

J. Goldemberg et al. "An End-Use Oriented Global Energy Strategy." *Annual Reviews of Energy 10* (1985): 613–688.

B. Keepin. "Review of Global Energy and Carbon Dioxide Projections." *Annual Reviews of Energy 11* (1986): 357–392.

A. Perry et al. "Energy Supply and Demand Implications of CO$_2$." *Energy 7* (1982): 991–1004.

Scenarios of Future CO$_2$ Buildup

J. Trabalka et al. "Atmospheric CO$_2$ Projections with Globally Averaged Carbon Cycle Models." In *The Changing Carbon Cycle: A Global Analysis,* J. Trabalka and D. Reichle, eds. New York: Springer-Verlag, 1986.

CLIMATIC RESPONSES TO CARBON DIOXIDE

Overview of Climate Modeling

R. Dickinson. "Modeling Changes Due to Carbon Dioxide Increases." In *Carbon Dioxide Review: 1982,* W. Clark, ed. New York: Oxford University Press, 1982.

J. Hansen et al. "Climate Impact of Increasing Atmospheric Carbon Dioxide." *Science 213* (1981): 957–966.

Radiative and Feedback Effects of CO$_2$ on Climate

V. Ramanathan. "The Role of Ocean-Atmosphere Interactions in the CO$_2$ Climate Problem." *Journal of Geophysical Research 38* (1981): 918–930.

J. Hansen et al. "Climate Sensitivity: Analysis of Feedback Mechanisms." In *Climate Processes and Climate Sensitivity,* J. Hansen and T. Takahashi, eds. Washington, D.C.: American Geophysical Union, 1984.

State of Climate Modeling

R. Dickinson. "Uncertainties of Estimates of Climatic Change." Position paper prepared for the workshop on "Developing Policies for Responding to Future Climate Change," Villach, Austria, October 1987. [Available from R. Dickinson, National Center for Atmospheric Research, P.O. Box 3000, Boulder, CO 80307.]

General Circulation Models and Modeling

S. Manabe and R. Wetherald. "Reduction in Summer Soil Wetness Induced by an Increase in Atmospheric Carbon Dioxide." *Science 232* (1986): 626–628.

S. Manabe, R. Wetherald, and R. Stouffer. "Summer Dryness Due to an Increase of Atmospheric CO_2 Concentration." *Climatic Change 3* (1981): 347–385.

S. Manabe and R. Stouffer. "Sensitivity of a Global Climate Model to an Increase of CO_2 Concentration in the Atmosphere." *Journal of Geophysical Research 85* (1980): 5529–5554.

C. Wilson and J. Mitchell. "A Doubled CO_2 Climate Sensitivity Experiment with a Global Climate Model Including a Simple Ocean." *Journal of Geophysical Research 92* (1987): 13315–13343.

M. Schlessinger. "General Circulation Model Simulations of CO_2-Induced Equilibrium Climate Change." In *Impact of Climatic Change on the Canadian Arctic,* H. French, ed. Toronto: Environment Canada, 1986.

W. L. Gates. "Problems and Prospects in Climate Modeling." In *Toward Understanding Climate Change,* R. Radok, ed. J. O. Fletcher Lectures on Problems and Prospects of Climate Analysis and Forecasting. Boulder: Westview Press, 1987.

Sea Level Response

R. R. Revelle. "Probable Future Changes in Sea Level Resulting from Increased Atmospheric Carbon Dioxide." In National Research Council, Carbon Dioxide Assessment Committee, *Changing Climate.* Washington, D.C.: National Academy Press, 1983.

J. G. Titus. "The Causes and Effects of Sea Level Rise." In *Effects of Changes in Stratospheric Ozone and Global Climate.* Vol. 1: *Overview,* J. Titus, ed. Washington, D.C.: U.S. Environmental Protection Agency, 1986.

R. Thomas. "Future Sea Level Rise and Its Early Detection by Satellite Remote Sensing." In *Effects of Change in Stratospheric Ozone and Global Climate.* Vol. 4: *Sea Level Rise,* J. Titus, ed. Washington, D.C.: U.S. Environmental Protection Agency, 1986.

J. Hecht. "America in Peril from the Sea." *New Scientist* (9 June 1988): 54–59.

G. deQ. Robin. "Projecting the Rise in Sea Level by Warming of the Atmosphere." In *The Greenhouse Effect, Climatic Change and Ecosystems,* B. Bolin et al., eds. Chichester: John Wiley and Sons, 1986.

T. Wigley and S. Raper. "Thermal Expansion of Sea Water Associated with Global Warming." *Nature 330* (1987): 127–131.

Impacts on the Cryosphere

R. Barry. "Possible CO_2-Induced Warming Effects on the Cryosphere." *Climatic Changes on a Yearly to Millennial Basis,* N. A. Morner and W. Karlena, eds. Dordrecht: D. Reidel Publishing Co., 1984.

Water Balance Response

T. Wigley and P. Jones. "Influence of Precipitation Changes and Direct CO_2 Effects on Streamflow." *Nature 314* (1985): 149–151.

R. Revelle and P. Waggoner. "Effects of a Carbon Dioxide-Induced Climatic Change on Water Supplies in the Western United States." In *Changing Climate,* Carbon Dioxide Assessment Committee, National Research Council. Washington, D.C.: National Academy Press, 1983.

Transient Response of Climate

J. Hansen et al. "Climate Response Times: Dependence on Climate Sensitivity and Ocean Mixing." *Science 229* (1985): 857–859.

K. Bryan et al. "Transient Climate Response to Increasing Atmospheric Carbon Dioxide." *Science 215* (1982): 56–58.

S. Thompson and S. Schneider. "Carbon Dioxide and Climate: The Importance of Realistic Geography in Estimating the Transient Temperature Response." *Science 217* (1982): 1031–1033.

W. Broecker. "Unpleasant Surprises in the Greenhouse?" *Nature 328* (1987): 123–126.

R. A. Kerr. "Is a Climate Jump in Store for Earth?" *Science 239* (1988): 259–260.

J. Gribbin. "Britain Shivers in the Global Greenhouse." *New Scientist* (9 June 1988): 42–43.

OTHER GASES

Overview of the Other Gas Question

V. Ramanathan et al. "Trace Gas Trends and Their Potential Role in Climate Change." *Journal of Geophysical Research 90* (1985): 5547–5566.

Past Emissions and Historic Rise in Concentrations

D. Wuebbles, M. MacCracken, and F. Luther. *A Proposed Reference Set of Scenarios for Radiatively Active Atmospheric Constituents.* DOE/NBB-0066. Washington, D.C.: U.S. Department of Energy, 1984.

G. Pearman et al. "Evidence of Changing Concentrations of Atmospheric CO_2, N_2O, and CH_4 from Air Bubbles in Antarctic Ice." *Nature 320* (1986): 248–250.

R. Rasmussen and M. Khalil. "Atmospheric Methane in the Recent and Ancient Atmospheres: Concentrations, Trends, and Interhemispheric Gradient." *Journal of Geophysical Research 89* (1984): 11599–11605.

Methane Sources and Budget

G. I. Pearman and P. J. Fraser. "Sources of Increased Methane." *Nature 332* (1988): 489–490.

D. R. Blake and F. S. Rowland. "Continued Worldwide Increase in Tropospheric Methane, 1978–1987." *Science 239* (1988): 1129–1131.

D. Ehhalt. "On the Rise: Methane in the Global Atmosphere." *Environment 27(10)* (1985): 8–33.

A. Holzapfel-Pschorn and W. Seiler. "Methane Emission During a Cultivation from an Italian Rice Paddy." *Journal of Geophysical Research 91* (1986): 11803–11814.

M. Khalil and R. Rasmussen. "Causes of Increasing Atmospheric Methane: Depletion of Hydroxyl Radicals and the Rise of Emissions." *Atmospheric Environment 19* (1985): 397–407.

H. Bingemer and P. Crutzen. "The Production of Methane from Solid Wastes." *Journal of Geophysical Research 92* (1987): 2181–2187.

Methane Response to Increasing Temperature

S. Hameed and R. Cess. "Impact of a Global Warming on Biospheric Sources of Methane and Its Climatic Consequences." *Tellus 35B* (1983): 1–7.

R. Revelle. "Methane Hydrates in Continental Slope Sediments and Increasing Atmospheric Carbon Dioxide." In *Changing Climate,* Carbon Dioxide Assessment Committee, National Research Council. Washington, D.C.: National Academy Press, 1983.

Nitrous Oxide Sources, Sinks, and Future Trends

W. Hao et al. "Sources of Atmospheric Nitrous Oxide from Combustion." *Journal of Geophysical Research 92* (1987): 3098–3104.

M. Kavanaugh. "Estimates of Future CO_2, N_2O, and NO_x Emissions from Energy Combustion." *Atmospheric Environment 21* (1987): 463–468.

Chlorofluorocarbons, Chlorocarbons, Fluorine Compounds: Present Emissions and Future Trends

J. Hammitt et al. "Future Emission Scenarios for Chemicals That May Deplete Stratospheric Ozone." *Nature 330* (1987): 711–716.

R. Dickenson and R. Cicerone. "Future Global Warming from Atmospheric Trace Gases." *Nature 319* (1986): 109–115. [See particularly the CFC emissions scenarios.]

P. Fabian et al. "CF_4 and C_2F_6 in the Atmosphere." *Journal of Geophysical Research 92* (1987): 9831–9835.

Carbon Monoxide, Nitric Oxide, and Ozone

J. Logan et al. "Tropospheric Chemistry: A Global Perspective." *Journal of Geophysical Research 86* (1981): 7210–7254.

M. Khalil and R. Rasmussen. "Carbon Monoxide in the Earth's Atmosphere: Indications of a Global Increase." *Nature 332* (1988): 242–245.

S. Hameed, R. Cess, and J. Logan. "Response of the Global Climate to Changes in Atmospheric Chemical Composition Due to Fossil Fuel Burning." *Journal of Geophysical Research 85* (1980): 7537–7545.

GLOBAL WARMING SIGNAL?

Overview

R. Kerr. "Is the Greenhouse Here?" *Science 239* (1988): 559–561.

Surface and Stratospheric Temperature Change

J. Hansen and S. Lebedeff. "Global Trends of Measured Surface Air Temperature." *Journal of Geophysical Research 92* (1987): 13345–13372.

K. Labitzke et al. "Long-Term Temperature Trends in the Stratosphere: Possible Influence of Anthropogenic Gases." *Geophysical Research Letters 13* (1986): 52–55.

T. Barnett. "Detection of Changes in the Global Troposphere Temperature Field Induced by Greenhouse Gases." *Journal of Geophysical Research 91* (1986): 6659–6667.

Other Indicators

J. Oerlemans. "Glaciers as Indicators of a Carbon Dioxide Warming." *Nature 320* (1986): 607–609.

R. Bradley et al. "Precipitation Fluctuations Over Northern Hemisphere Land Areas Since the Mid-19th Century." *Science 237* (1987): 171–175.

IMPACTS OF GLOBAL WARMING

Unmanaged Biosphere

A. Solomon and D. West. "Atmospheric Carbon Dioxide Change: Agent of Future Forest Growth or Decline?" In *Effects of Changes in Stratospheric Ozone and Global Climate.* Vol. 3: *Climate Change,* J. Titus, ed. Washington, D.C.: U.S. Environmental Protection Agency, 1986.

R. Peters and J. Darling. "The Greenhouse Effect and Nature Reserves." *Bioscience 35* (1985): 707–717.

R. Peters. "Effects of Global Warming on Biological Diversity: An Overview." In *Preparing for Climatic Change.* Washington, D.C.: Government Institutes, Inc., 1987.

C. Harrington. "The Impact of Changing Climate on Some Vertebrates in the Canadian Arctic." In *Impact of Climatic Change on the Canadian Arctic,* H. French, ed. Toronto: Environment Canada, 1986.

Agriculture

C. Rosenzweig. "Potential CO_2-Induced Climate Effects on North American Wheat-Producing Regions." *Climatic Change 7* (1985): 367–389.

R. Stewart. "Climatic Change—Implications for the Prairies." In *Effects of Changes in Stratospheric Ozone and Global Climate.* Vol. 3: *Climate Change,* J. Titus, ed. Washington, D.C.: U.S. Environmental Protection Agency, 1986.

E. Cooter. "An Assessment of the Potential Economic Impacts of Climate Change in Oklahoma." In *Effects of Changes in Stratospheric Ozone and Global Climate.* Vol. 3: *Climate Change,* J. Titus, ed. Washington, D.C.: U.S. Environmental Protection Agency, 1986.

T. Blasing and A. Solomon. *Response of the North American Corn Belt to Climate Warming.* DOE/NBB-0040. Washington, D.C.: U.S. Department of Energy, 1983.

P. Waggoner. "Agriculture and Carbon Dioxide." *American Scientist 72* (1984): 179–184.

Sea Level Rise

P. Vellinga. "Sea Level Rise, Consequences and Policies." Position paper prepared for the workshop on "Developing Policies for Responding to Future Climate Change," Villach, Austria, October 1987. [Available from P. Vellinga, Delft Hydraulics, P.O. Box 152, 8300 AD EMMELOORD, Netherlands.]

R. Park, T. Armentano, and C. L. Cloonan. "Predicting the Effects of Sea Level Rise on Coastal Wetlands." In *Effects of Changes in Stratospheric Ozone and Global Climate.* Vol. 4: *Sea Level Rise,* J. Titus, ed. Washington, D.C.: U.S. Environmental Protection Agency, 1986.

S. Leatherman. "Effects of Sea Level Rise on Beaches and Coastal Wetlands." Paper presented at Climate Institute Symposium on Climate Change Impacts on Wildlife, Washington, D.C., January 1988. [Available from S. P. Leatherman, Laboratory for Coastal Research, University of Maryland, 1175 Lefrak Hall, College Park, MD 20742.]

Impacts Stemming from Changes in Stream Flow and Lake Levels

P. Gleick. "Regional Water Resources and Global Climatic Change." In *Effects of Changes in Stratospheric Ozone and Global Climate.* Vol. 3: *Climate Change,* J. Titus, ed. Washington, D.C.: U.S. Environmental Protection Agency, 1986.

S. Cohen. "Impacts of CO_2-Induced Climatic Change on Water Resources in the Great Lakes Basin." *Climatic Change 8* (1986): 135–153.

S. Cohen. "Climatic Change, Population Growth, and Their Effects on Great Lakes Water Supplies." *Professional Geographer 38* (1986): 317–323.

Impacts in High Latitudes

G. McKay and W. Baker. "Socio-Economic Implications of Climate Change in the Canadian Arctic." In *Impact of Climatic Change on the Canadian Arctic,* H. French, ed. Toronto: Environment Canada, 1986.

PERSPECTIVES ON THE POLICY RESPONSE

Economic Paradigms

L. Lave. "Mitigating Strategies for Carbon Dioxide Problems." *American Economic Review 72* (1982): 257–261.

R. D'Arge, W. Schulze, and D. Brookshire. "Carbon Dioxide and Intergenerational Choice." *American Economic Review 72* (1982): 251–256.

W. Nordhaus. "How Fast Should We Graze the Global Commons?" *American Economic Review 72* (1982): 242–246.

T. Schelling. "Climatic Change: Implications for Welfare and Policy." In *Changing Climate,* Carbon Dioxide Assessment Committee, National Research Council. Washington, D.C.: National Academy Press, 1983.

P. G. Brown. "Policy Analysis, Welfare Economics, and the Greenhouse Effect." *Journal of Policy Analysis and Management 7(3)* (1988): 471–475.

Muddling Through

M. Glantz. "A Political View of CO_2." *Nature 280* (1979): 189–190.

Anticipatory Adaptation

I. Mintzer. "Living in a Warmer World: Challenges for Policy Analysis and Management." *Journal of Policy Analysis and Management 7(3)* (1988): 445–459.

L. B. Lave. "The Greenhouse Effect: What Government Actions Are Needed?" *Journal of Policy Analysis and Management 7(3)* (1988): 460–470.

Limit Global Heating

G. M. Woodwell. "Global Warming and What We Can Do About It." *Amicus Journal* (Fall 1986): 8–12.

W. R. Moomaw. "Proposed Near-Term Congressional Options for Responding to Global Climate Change." Mimeo. Washington, D.C.: World Resources Institute, 1988.

MEANS TO LIMIT GLOBAL HEATING

Energy Policy

A. Perry. "Possible Changes in Future Use of Fossil Fuels to Limit Environmental Effects." In *The Changing Carbon Cycle: A Global Analysis,* J. Trabalka and D. Reichle, eds. New York: Springer-Verlag, 1986.

D. Rose, M. Miller, and C. Agnew. "Reducing the Problem of Global Warming." *Technology Review* (May/June 1984): 49–58.

J. Goldemberg et al. "Basic Needs and Much More with One Kilowatt per Capita." *Ambio 14(4–5)* (1985): 190–200.

W. U. Chandler. *Energy Productivity: Key to Environmental Protection and Economic Progress.* Worldwatch Paper 63. Washington, D.C.: Worldwatch Institute, 1985.

T. B. Johansson and R. H. Williams. "Energy Conservation in the Global Context." *Energy 12(10/11)* (1987): 907–919.

R. H. Williams. "A Low Energy Future for the United States." *Energy 12(10/11)* (1987): 929–944.

W. Bach. "Carbon Dioxide/Climate Threat: Fate or Forbearance?" In *Carbon Dioxide: Current Views and Developments in Energy/Climate Research,* W. Bach et al., eds. Dordrecht: D. Reidel Publishing Co., 1983.

Carbon Dioxide Scrubbers

M. Steinbert, H. Chang, and F. Horn. *A Systems Study for the Removal, Recovery, and Disposal of Carbon Dioxide from Fossil Fuel Power Plants in the U.S.* DOE/CH/00016–2. Washington, D.C.: U.S. Department of Energy, 1984.

B. Louks. "CO_2 Production in Gasification-Combined-Cycle Plants." *EPRI Journal* (October/November 1987): 52–54.

G. MacDonald. Testimony, U.S. Senate Committee on Energy and Natural Resources, 9–10 November 1987, Senate Hearing 100–461, at pages 248 and 283–286. [See also G. M. Hildy, same hearings, at page 248, and J. F. Decker (U.S. Department of Energy), same hearings, at pages 275–276.]

Seeding Stratosphere with Particulates to Induce Global Cooling

W. W. Kellogg and S. H. Schneider. "Climate Stabilization: For Better or for Worse?" *Science 186* (1974): 1163–1172.

S. H. Schneider. "The Greenhouse Effect: What We Can or Should Do About It." In *Preparing for Climate Change.* Washington, D.C.: Government Institutes, Inc., 1987.

Decreasing Atmospheric CO_2 by Reforestation

F. Dyson. "Can We Control the Carbon Dioxide in the Atmosphere?" *Energy 2* (1977): 287–291.

Biospheric Eutrophication

B. J. Peterson and J. M. Melillo. "The Potential Storage of Carbon Caused by Eutrophication of the Biosphere." *Tellus 37B* (1985): 117–127.

Emissions Control of Other Gases

Office of Technology Assessment. "An Analysis of the Montreal Protocol on Substances That Deplete the Ozone Layer." Staff paper. Washington, D.C.: OTA, 1987.

J. A. Laurmann. "Emissions Control and Reduction." Position paper prepared for the workshop on "Developing Policies for Responding to Future Climate Change," Villach, Austria, October 1987. [Available from J. A. Laurmann, Gas Research Institute, 8600 W. Bryn Mawr Ave., Chicago, IL 60656.]

DECLARATIONS AND CONFERENCE STATEMENTS

1985 Villach Conference Statement

World Climate Program. *Report of the International Conference on the Assessment of the Role of Carbon Dioxide and of Other Greenhouse Gases in Climate Variations and Associated Impacts.* Report of an international conference held at Villach, Austria, 9–15 October 1985. World Meteorological Organization, WMO-661.

1987 Villach Conference Statement

World Climate Program. *Developing Policies for Responding to Climatic Change.* A summary of the discussions and recommendations of the workshops held in Villach, Austria (28 September–2 October 1987), and Bellagio, Italy (9–13 November 1987), J. Jaeger, ed. World Meteorological Organization, WCIP-1 (April 1988).

1987 Joint U.S.–Soviet Summit Statement

Text of the joint U.S.–USSR Summit Statement, 7–10 December 1987, Washington, D.C., Paragraph IV: Bilateral Affairs.

1988 Toronto Conference Statements

Conference Statement: "The Changing Atmosphere: Implications for Global Security." Toronto, 27–30 June 1988.

Nongovernmental Organization Statement: "Escaping the Heat Trap: An NGO Statement of Policies to Prevent Climate Change."

FOR FURTHER INFORMATION

Many government agencies and nongovernmental organizations have greenhouse heating and climatic change programs, and the list grows rapidly. Included among them are the following.

FEDERAL AGENCIES

U.S. Department of Energy
Office of Energy Research
Office of Basic Energy Sciences
Carbon Dioxide Research Division
Washington, D.C. 20585

U.S. Environmental Protection Agency
Office of Policy Planning and Evaluation
401 M Street SW
Washington, D.C. 20460

NEWSLETTERS

CDIAC Communications
  Carbon Dioxide Information Analysis Center
  Building 2001, MS-050
  Oak Ridge National Laboratory
  P.O. Box 2008
  Oak Ridge, TN 37831

Climate Alert
  Climate Institute
  316 Pennsylvania Avenue SE, Suite 403
  Washington, D.C. 20003

CO₂/Climate Report
Climate Program Office
Environment Canada
4905 Dufferin Street
Downsview, Ontario M3H 5T4
Canada

Greenhouse Effect Report
Business Publishers Inc.
951 Pershing Drive
Silver Springs, MD 20910

NONGOVERNMENTAL ORGANIZATIONS

Beijer Institute
P.O. Box 50005
S-104 05 Stockholm
Sweden

Climate Institute
316 Pennsylvania Avenue SE, Suite 403
Washington, D.C. 20003

Electric Power Research Institute
P.O. Box 10412
Palo Alto, CA 94303

Environmental Defense Fund
257 Park Avenue South
New York, NY 10010

Environmental Policy Institute
218 D Street SE
Washington, D.C. 20003

Friends of the Earth
530 7th Street SE
Washington, D.C. 20003

National Audubon Society
950 Third Avenue
New York, NY 10022

Natural Resources Defense Council
1350 New York Avenue NW
Washington, D.C. 20005

Renew America
1001 Connecticut Avenue NW
Washington, D.C. 20036

Sierra Club
730 Polk Street
San Francisco, CA 94109

World Resources Institute
1735 New York Avenue NW
Washington, D.C. 20006

World Wildlife Fund
1250 24th Street NW
Washington, D.C. 20037

INDEX

341

ABOUT THE AUTHORS

AUTHORS OF CHAPTERS WRITTEN FOR THIS BOOK

Dean Edwin Abrahamson is professor of public affairs at the University of Minnesota's Hubert H. Humphrey Institute of Public Affairs, cochair of the university's Council on Environmental Quality, an adjunct professor at the University of Iceland, and a member of the board of the Natural Resources Defense Council. He has degrees in physics, medicine, and biology and has held a variety of jobs in industry, government, and academia. Since the late 1960s his primary work has been with environmental and public health implications of energy policies, and for the last several years it has focused on the policy implications of the greenhouse effect.

Peter Ciborowski is a research fellow at the Hubert H. Humphrey Institute of Public Affairs, University of Minnesota, working on global warming. He took his undergraduate degree from the State University of New York at Albany and a master's degree at the University of Minnesota. He is presently pursuing a Ph.D. and has published and spoken widely on atmospheric issues. His research interests include energy policy impacts on the greenhouse gases, strategies for limiting chlorofluorocarbon and fossil fuel usage, and strategic responses to the global warming question.

John Firor is director of the Advanced Study Program at the National Center for Atmospheric Research (NCAR) in Boulder, Colorado. He came to NCAR in 1961 as director of the High Altitude Observatory of NCAR; he was made director of NCAR in 1968 and assumed his present position in 1980. Earlier he had served as a staff member, in radio astronomy, of the Department of Terrestrial Magnetism of the Carnegie Institution of Washington. He holds a B.S. in physics from the Georgia Institute of Technology and a Ph.D. in physics from the University of Chicago. He is a member of a number of professional societies and in 1972 was elected a fellow of the American Meteorological Society. While on the staff of NCAR he has also served as a professor-adjoint at the University of Colorado, a visiting professor at Cal Tech, a visiting scholar at Resources for the Future, and a senior fellow at the Hubert H. Humphrey Institute of Public Affairs at the University of Minnesota. He has been a trustee of the Environmental Defense Fund since 1974 and was chairman from 1975 to 1980. He is a founding trustee of the World Resources Institute and a trustee of the International Federation of Institutes of Advanced Study. He has written articles in scientific journals, books, and popular publications on cosmic rays, radio sources in the universe, the sun's atmosphere, solar flares, the global climate, absorption of carbon dioxide by the oceans, climate change, acid rain, and depletion of stratospheric ozone.

Rafe Pomerance is senior associate for policy affairs at the World Resources Institute, Washington, D.C. He has a B.A. from Cornell University. He is vice-chairman of the board of the League of Conservation Voters and a member of the boards of the Climate Institute and the National Clean Air Coalition. He was associate legislative director for Friends of the Earth and coordinator of the National Clean Air Coalition from 1975 to 1978, the director of Friends of the Earth from 1978 to 1980, and its president and chief executive officer from 1980 to 1984. He has been a member of the U.S. Department of Energy Advisory Committee on Carbon Dioxide, a member of the U.S. delegation to the UNEP Governing Council, and an observer during the Geneva/UNEP negotiations to protect the ozone layer in 1986 and 1987.

Timothy E. Wirth is a U.S. senator from the state of Colorado.

George M. Woodwell is an ecologist with broad interests in global environmental issues and policies. Prior to founding the Woods Hole

Research Center, where he is director, he was founder, director, and distinguished scientist at the Ecosystems Center of the Marine Biological Laboratory in Woods Hole and a senior scientist at the Brookhaven National Laboratories. Dr. Woodwell holds an adjunct faculty position in ecology at Yale University. He is a founder and member of the board of trustees of the Environmental Defense Fund, a founder and currently vice-chairman of the board of the Natural Resources Defense Council, former chairman of the board of trustees of the World Wildlife Fund, a founding and current trustee of the World Resources Institute, and former president of the Ecological Society of America. Dr. Woodwell is the author of over 200 major papers and books in ecology. He holds a doctorate in botany from Duke University and has received three honorary degrees.

AUTHORS OF PREVIOUSLY PUBLISHED CHAPTERS

Donald R. Blake is professor of chemistry at the University of California, Irvine.

Wallace S. Broecker is professor of geochemistry, Lamont-Doherty Geological Observatory, Columbia University, New York.

Ralph Cavanagh is codirector of the energy program of the Natural Resources Defense Council, San Francisco.

Ralph Cicerone is director of the Atmospheric Chemistry Division at the National Center for Atmospheric Research, Boulder, Colorado.

David Goldstein is codirector of the energy program of the Natural Resources Defense Council, San Francisco.

James E. Hansen is director of the NASA Goddard Institute for Space Studies, New York.

Jill Jaeger is a climate scientist who served as consultant to the Beijer Institute in organizing the 1987 Villach-Bellagio workshops.

Gordon MacDonald is vice-president and chief scientist at the MITRE Corporation, McLean, Virginia.

Syukuro Manabe is a climatologist at the Geophysical Fluid Dynamics Laboratory, National Oceanic and Atmospheric Administration, Princeton, New Jersey.

William R. Moomaw is director of the Climate, Energy, and Pollution Program, World Resources Institute, Washington, D.C.

Robert L. Peters is a biologist with the World Wildlife Fund, Washington, D.C.

V. Ramanathan is professor of geophysical sciences, University of Chicago.

Roger R. Revelle is professor of science and public policy at the University of California, San Diego.

James G. Titus is a senior scientist with the U.S. Environmental Protection Agency, Washington, D.C.

Paul E. Waggoner is director of the Connecticut Agricultural Experiment Station.

Robert Watson is a research associate in the energy program of the Natural Resources Defense Council, San Francisco.

ALSO AVAILABLE FROM ISLAND PRESS

Americans Outdoors: The Report of the President's Commission
The Legacy, The Challenge, with case studies
Foreword by William K. Reilly
1987, 426 pp., appendices, case studies, charts
Paper: $24.95 ISBN 0-933280-36-X

Crossroads: Environmental Priorities for the Future
Edited by Peter Borrelli
1988, 352 pp., index
Cloth: $29.95 ISBN: 0-933280-68-8
Paper: $17.95 ISBN: 0-933280-67-X

Deserts on the March
By Paul B. Sears
New Introduction by Gus Speth, President, World Resources Institute
Conservation Classic Edition 1988
256 pp., illustrations
Cloth: $29.95 ISBN 0-933280-46-7
Paper: $19.95 ISBN 0-933280-90-04

Down by the River: The Impact of Federal Water Projects and Policies on Biodiversity
By Constance E. Hunt with Verne Huser
In cooperation with The National Wildlife Federation
1988, 256 pp., illustrations, glossary, index, bibliography
Cloth: $34.95 ISBN: 0-933280-48-3
Paper: $22.95 ISBN: 0-933280-47-5

Forest and the Trees: A Guide to Excellent Forestry
By Gordon Robinson, Introduction by Michael McCloskey
1988, 272 pp., indexes, appendices, glossary, tables, figures
Cloth: $34.95 ISBN: 0-933280-41-6
Paper: $19.95 ISBN: 0-933280-40-8

Holistic Resource Management
By Allan Savory
Center for Holistic Resource Management
1988, 512 pp., plates, diagrams, references, notes, index
Cloth: $39.95 ISBN: 0-933280-62-9
Paper: $24.95 ISBN: 0-933280-61-0

Land and Resource Planning in the National Forests
By Charles F. Wilkinson and H. Michael Anderson, Introduction by Arnold Bolle
1987, 400 pp., index
Paper: $19.95 ISBN: 0-933280-38-6

Last Stand of the Red Spruce
By Robert A. Mello
Introduction by Senator Patrick J. Leahy
In cooperation with the Natural Resources Defense Council
1987, 208 pp.
Paper: $14.95 ISBN: 0-933280-37-8

The Poisoned Well: New Strategies for Groundwater Protection
Sierra Club Legal Defense Fund
1988, 225 pp., index, glossary, charts, appendices, bibliography
Cloth: $31.95 ISBN: 0-933280-56-4
Paper: $19.95 ISBN: 0-933280-55-6

Reopening the Western Frontier
From High Country News
1989, 300 pp., illustrations, photographs, index
Cloth: $24.95 ISBN: 1-55963-011-6
Paper: $15.95 ISBN: 1-55963-010-8

Saving the Tropical Forests
By Judith Gradwohl and Russell Greenberg
Preface by Michael H. Robinson
Smithsonian Institution
1988, 207 pp., index, tables, illustrations, notes, bibliography
Cloth: $24.95 ISBN: 0-933280-81-5

For a complete catalog of Island Press publications, please write:
Island Press, Box 7, Covelo, CA 95428.